Specifications and Criteria for Biochemical Compounds

Supplement: Biogenic Amines and Related Compounds

Specifications and Criteria for Biochemical Compounds

Supplement: Biogenic Amines and Related Compounds

Subcommittee on Biogenic Amines
and Related Compounds
Committee on Specifications and Criteria
for Biochemical Compounds
Assembly of Mathematical and Physical Sciences
National Research Council

NATIONAL ACADEMY OF SCIENCES
Washington, D.C. 1977

NOTICE: The project that is the subject of this report was approved by the Governing Board of the National Research Council, whose members are drawn from the Councils of the National Academy of Sciences, the National Academy of Engineering, and the Institute of Medicine. The members of the Committee responsible for the report were chosen for their special competences and with regard for appropriate balance.

This report has been reviewed by a group other than the authors according to procedures approved by a Report Review Committee consisting of members of the National Academy of Sciences, the National Academy of Engineering, and the Institute of Medicine.

Support for this project was provided by the National Institutes of Health under Contract No. NO1-GM-5-2153.

ISBN 0-309-02601-6

Library of Congress Catalog Card Number 77-72355

Available from
Printing and Publishing Office
National Academy of Sciences
2101 Constitution Avenue, N.W.
Washington, D.C. 20418

Printed in the United States of America

Contents

	Code Number	Page
Introduction		1
Acknowledgments		5

Data Sheets

	Code Number	Page
D-(−)-Amphetamine sulfate	BA-1	7
L-(+)-Amphetamine sulfate	BA-2	7
3,4-Dimethoxyphenethylamine hydrochloride	BA-3	7
N,N-Dimethyltryptamine	BA-4	8
Dopamine hydrochloride	BA-5	8
D-(−)-Epinephrine	BA-6	9
D-(−)-Epinephrine (+)-tartrate	BA-7	9
L-(+)-Epinephrine (−)-tartrate	BA-8	10
Gramine	BA-9	10
Histamine dihydrochloride	BA-10	11
Hordenine sulfate dihydrate	BA-11	11
Melatonin	BA-12	12
Mescaline hydrochloride	BA-13	12
DL-Metanephrine hydrochloride	BA-14	12
5-Methoxytryptamine	BA-15	13
D-(−)-Norepinephrine	BA-16	13
DL-Norepinephrine	BA-17	14
DL-Norepinephrine hydrochloride	BA-18	14
D-(−)-Norepinephrine (+)-tartrate monohydrate	BA-19	15
DL-Normetanephrine hydrochloride	BA-20	15
DL-Octopamine hydrochloride	BA-21	16
Phenethylamine hydrochloride	BA-22	16
Serotonin creatinine sulfate monohydrate	BA-23	17
Serotonin hydrogen oxalate	BA-24	17
DL-Synephrine hydrochloride	BA-25	18
DL-Synephrine (+)-tartrate	BA-26	18
Tryptamine	BA-27	19
Tryptamine hydrochloride	BA-28	19
Tyramine	BA-29	20
Tyramine hydrochloride	BA-30	20

Introduction

Guggenheim[1] defined biogenic amines as "organic bases of low molecular weight which arise in consequence of metabolic processes in animals, plants and micro-organisms. They comprise aliphatic, alicyclic and simple heterocyclic compounds, and appear in cellular metabolism as intermediary or catabolic products of varied physiological importance." His list of structural types included alkylamines, alkenylamines, aminoalkyl alcohols, diamines, guanidine, imidazoles, indoles, phenethylamines, betaines, and other materials of as yet unknown constitution. Amino acids, amino sugars, nucleotides, porphyrins, and plant-derived alkaloids generally are excluded from this class of biochemical compounds.

The Subcommittee on Biogenic Amines and Related Compounds has reviewed only compounds that are commercially available. Among the compounds are some of the most common biogenic amines or optical antipodes thereof occurring naturally in humans and animals. Related compounds, such as gramine, a plant alkaloid, and two amphetamines (non-biogenic), are included among these amines. These materials are of general interest to those engaged in biochemical research and in clinical studies of normal and pathological states. They were obtained as the highest grade substances available from normal stocks of manufacturers or suppliers. All of these compounds are synthetic.

It is important that biological and physiological chemists use appropriate means to establish the acceptability of the compounds they select for study. To this end, the Subcommittee presents, in data sheet form, uncomplicated analytical processes, related data, and references for each particular biogenic amine. This information can be used to establish the suitability of the products for their intended experimental purpose. Laboratories using a particular amine extensively should establish a primary standard for comparisons between various lots of the compound they obtain.

In the data sheets, optical isomers and salt forms have been treated separately. Almost all of the physical data were determined in the laboratories of members of the Subcommittee. Terminology and methodology generally applied were taken from the parent publication, *Specifications and Criteria for Biochemical Compounds*, Third Edition (1972),[2] and the *United States Pharmacopeia*, XIX Revision (1975).[3]

Data Sheet Format

Indexing terms, the structural formula, and the calculated formula weight applicable to a particular biogenic amine appear at the top of each data sheet. Under each compound number, the first line is the common or nonproprietary name of the compound chosen by the Subcommittee. The second line is the IUPAC systematic chemical name, frequently the *Chemical Abstracts Index* name used through 1971. The third line is the current *C.A. Index* name. This third name is needed for entry into *C.A. Indexes* since 1972, but is not necessarily the most useful chemical name. The fourth line lists other synonyms and trade names; the latter are indicated by capital initial letters. The archaic *dl* optical rotation designations are replaced by (±), and configurational assignments are given as D/L forms. The *C.A.* name includes the configuration in the Ingold–Prelog R/S system. Water content is given as a calculated value when water of crystallization is referred to. Following this informational section are criteria profiles dealing with the physical description of the compound, identification, purity, source, conditions for storage and handling, and references. Results of tests for residue on ignition, water content, heavy metals, and loss on drying observed for the substances are included among the introductory information.

Description

The appearance or physical description are given for the compounds as received from commercial suppliers. Melting points are expressed as ranges obtained by using a Thomas–Hoover capillary melting point apparatus with an ASTM E1 thermometer. The accuracy of this apparatus was checked using six compounds of specified melting ranges in accord with the *U.S. Pharmacopeia*.[3] Actual melting points for the reference standards were all within 1° of the reported ranges from 80° to 235°. It is generally recognized that the melting point is often useful in confirming the identity of biogenic amines and, less often, as a criterion of purity for this class of compounds. The reservation about relating melting point to purity is founded, in part, on observed thermal instabilities, zwitterionic possibilities, and the occurrence of variations in optical forms for some biogenic amines. The solubility and stability information was taken for the most part from various compendia and handbooks, such as *The Merck Index*.[4] Some notations on substance stability are from the personal experience of members of the Subcommittee.

The absorption spectrum, native fluorescence spectrum, and the fluorescence of reaction products are often useful in characterizing individual biogenic amines. Procedures for achieving significant and valid spectrophotometric and fluorometric results are discussed in ref. 3, p. 662. The anionic species in salt forms are identified by relatively specific tests, as described under *Identification Tests* in the *United States Pharmacopeia*,[3] and as noted in the data sheets.

The ultraviolet absorption spectra of the compounds predried to constant weight over phosphorus pentoxide were recorded for solutions prepared in 0.1 M hydrochloric acid with Zeiss and Cary 14 spectrophotometers. The wavelength accuracy of the instruments was checked with a mercury lamp.[5] Other calibration procedures may be equally useful.[6,7] Instrument detector response was monitored in absorbance units with reference standards prepared at the Sterling Winthrop Research Institute. In brief, absorbance values for standards were established subsequent to an earlier study that utilized a pairing experiment involving 33 spectrophotometers. Statistical pairing experiments have been described by Youden.[8] In this application, absorbance values were recorded for the reference standard, mepivacaine hydrochloride, in water at two concentration levels in a 10-mm rectangular quartz cuvette at 263 nm. Instruments were considered suitable for estimating molecular extinction values for biogenic amines if they gave values within 2 standard deviations of the established average absorbance values for the standard mepivacaine preparations. Other reference materials for checking detector response have been described.[7,9] Potassium dichromate has been used as a standard to check the performance of ultraviolet photometers.[10] Spectral absorbance standards are available as ultra-high-purity materials (Ultrex grade) from J. T. Baker Chemical Company. The values given in the data sheets are maximum wavelengths of absorption λ_{max}, shoulders λ_{sh}, and molar absorptivity ϵ.

The native excitation and emission spectra are useful criteria for identification of many biogenic amines in solution. Quantitative data may also be obtained by comparing the relative fluorescence of a standard solution of a biogenic amine with that of a sample solution of unknown concentration at appropriate fixed excitation and emission wavelengths. The methods used here for fluorescence measurements are described by Udenfriend.[11]

The fluorescence emission maxima are corrected values obtained with an Amicon–Keirs fluorescence spectrometer equipped with a linear/log Varicord Model 43 recorder. A xenon light source was used, and solutions were examined in a 10-mm quartz cuvette at ambient room temperature, ~22°. Wavelength accuracy, both for excitation and emission, was determined with a quartz mercury lamp by the method described in Appendix II of the text by Udenfriend.[11] The characteristics of the monochromator dials limited the accuracy of the reading to ±3 nm. Maximum instrument efficiency was established using the methodology described in Appendixes II and III of Udenfriend.[11] The minimum detectable concentration of quinine sulfate was estimated at ~50 ppb with this instrument.

Standardized procedures were followed in preparing samples for spectral determination. Freshly distilled water, or buffers prepared each week, were used as solvents. Buffers were made using reagent-grade chemicals. The spectra of all solvents were determined; none exhibited peaks in the area of absorption or emission of the biogenic amines tested. The concentration of biogenic amine in water or buffer solution was 1 μg/ml for all substances examined, except as otherwise noted. The values listed in the data sheets are the maximum wavelengths of excitation λ_{ex} and emission λ_{em}.

When characteristic fluorescent reaction products of various amines were produced, a reagent blank without amine was carried through the procedure and used in place of solvent to determine background fluorescence.

Homogeneity

Thin-layer chromatography (TLC) of biogenic amines is a convenient method for detecting organic chemical impurities. Impurities in the range of 1% (and higher) generally can be detected by this method. The usefulness of TLC in establishing drug purity profiles has been discussed by Grady *et al.*[12]

Chromatographic systems differing in adsorbent or developing solvent were used in conjunction with several methods of detection. Two separate thin-layer chromatograms were obtained using, respectively, the adsorbents silica gel F-254 and cellulose F. Usually a 4-μl aliquot of a solution of the compound in an appropriate solvent (25 mg/5 ml) was spotted on the plates, which were then developed with a suitable solvent to a height of 10–11 cm. For the phenolic amines, especially those having catechol structure, the plates may be sprayed with a 1% solution of sodium bisulfite prior to activation. The R_f values on the data sheets were obtained using precoated Merck plates without the bisulfite spray.

Detection Reagents

Various reagents and methods were used to detect the biogenic amines on the TLC plates.

1. Ninhydrin Reagent was prepared by dissolving 0.3 g of ninhydrin (1,2,4-indanetrione monohydrate) in 100 ml of *n*-butanol, adding 3 ml of glacial acetic acid, and mixing.

2. Folin–Ciocalteau Reagent, referred to in the data sheets as Folin's Reagent, was prepared according to ref. 3, p. 765 and p. 181. It is now commercially available. To detect amines, the TLC plates are sprayed with this reagent followed by a spray of 10% sodium carbonate.

3. Ehrlich's Reagent can be prepared in several ways. For tryptamine and serotonin, the plates are sprayed with a 1% solution of *p*-dimethylaminobenzaldehyde in 96% aqueous ethanol. After being sprayed, the plates are exposed to HCl vapor for 3–5 minutes. For indole compounds generally, 1 g of the Reagent is dissolved in 50 ml of concentrated HCl, and 50 ml of ethanol is added. Plates sprayed with this Reagent can be fixed further by blowing the vapors of aqua regia over the chromatograms. A modified Ehrlich's Reagent has been reported for the quantitative measurement of indole compounds.[13]

4. Iodine vapor: the developed TLC plates are dried in air and exposed to iodine vapor in a closed jar.

5. Iodine in ethanol: the developed TLC plates are sprayed with a dilute solution of iodine in ethanol.

6. Iodoplatinate Reagent is made by dissolving 300 mg of platinic chloride in 100 ml of water and adding 100 ml of potassium iodide solution containing 6 g of KI. The solution is stable when mixed and stored in a brown glass container.

7. Fluorescence: a number of biogenic amine compounds fluoresce in ultraviolet light.

These and other methods for the detection of biogenic amines on paper and thin-layer chromatograms are described by Weil-Malherbe.[14]

TABLE 1

Compound	Heavy[a] metals (ppm)	Residue[b] on ignition (%)	Water[c] content (%)	Loss on[d] drying (%)
D-(−)-Amphetamine sulfate	<10	0.10	0.20	0.1
L-(+)-Amphetamine sulfate	<10	0.00	0.22	0.2
3,4-Dimethoxyphenethylamine hydrochloride	<10	0.00	0.77	0.3
N,*N*-Dimethyltryptamine	<10	0.17	0.65	0.2
Dopamine hydrochloride	<10	0.12	0.16	0.4
D-(−)-Epinephrine	<10	0.27	1.2	1.7
D-(−)-Epinephrine (+)-tartrate	<10	0.00	0.16	0.5
L-(+)-Epinephrine (−)-tartrate	<10	0.00	1.2	0.9
Gramine	<10	0.24	0.17	0.2
Histamine dihydrochloride	<10	0.00	0.16	0.2
Hordenine sulfate dihydrate	<10	1.81	7.0	6.8
Melatonin	<10	0.17	0.00	0.8
Mescaline hydrochloride	<10	0.10	0.68	0.4
DL-Metanephrine hydrochloride	<10	0.00	0.00	1.9
5-Methoxytryptamine	<10	0.00	0.20	0.7
D-(−)-Norepinephrine	<10	0.00	1.2	0.4
DL-Norepinephrine	<10	0.05	0.96	0.4
DL-Norepinephrine hydrochloride	<10	0.00	0.22	0.2
D-(−)-Norepinephrine (+)-tartrate monohydrate	<10	0.00	5.55	0.4
DL-Normetanephrine hydrochloride	<10	0.00	0.54	0.2
DL-Octopamine hydrochloride	<10	0.10	0.30	0.2
Phenethylamine hydrochloride	<10	0.00	0.15	0.5
Serotonin creatinine sulfate monohydrate	<10	0.20	4.7	4.6
Serotonin hydrogen oxalate	<10	0.22	0.70	0.6
DL-Synephrine hydrochloride	<10	0.07	0.16	0.4
DL-Synephrine (+)-tartrate	<10	0.25	0.00	0.4
Tryptamine	<10	0.09	0.20	0.7
Tryptamine hydrochloride	<10	0.17	0.00	0.5
Tyramine	<10	0.00	0.33	0.5
Tyramine hydrochloride	<10	0.00	0.12	0.6

[a] *U.S. Pharmacopeia*, XIX, p. 619—given as parts by weight of lead per million parts of the test substance.
[b] *U.S. Pharmacopeia*, XIX, p. 620.
[c] *U.S. Pharmacopeia*, XIX, p. 668, Karl Fischer titrimetric method.
[d] Dried to constant weight over P_2O_5 *in vacuo* at room temperature.

Specific Rotation

Several of the biogenic amine compounds examined are optically active. The specific rotation is of value in establishing the identity of the total salt, and the optical purity of the parent amine, if the anion plays no principal part in the gross rotation. A discussion of the method for determining optical rotation is contained in ref. 3, p. 652, which contains practical suggestions for maintaining accuracy and precision in optical rotation measurements. The equation for calculating specific rotation is given in the parent volume, ref. 2, p. 2.

It is important to characterize the portion of the compound that imparts the physiological properties of interest. In this respect, for example, one is usually concerned primarily with the purity of D-(−)-norepinephrine and only secondarily with the tartaric moiety of its (+)-tartrate monohydrate. A specific rotation of −10° to −11° for the tartrate does not really provide a great deal of assurance about the state of optical purity of D-(−)-norepinephrine. One can focus on the optical purity of norepinephrine by measuring the free base derived from the tartrate ($[\alpha]_D$ −38°), or better yet, the N-acetyl-3,4-di-O-acetyl derivative can be prepared quantitatively and its optical purity ($[\alpha]_D$ −81.3°) measured. The procedure used to measure the optical purity of D-(−)-norepinephrine and (+)- and (−)-epinephrine is described in ref. 3, pp. 170 and 280, and can be readily adapted to a solid sample.

Heavy Metals, Residue on Ignition, Water Content, and Loss of Weight on Drying

The Subcommittee determined these values for the products obtained from commercial sources. The data obtained on single samples were generally within acceptable limits. Somewhat high values of water content were obtained by the Karl Fischer titrimetric method for several compounds. In some instances, the values for water content were comparable to the less specific values for loss on drying. The data are given in Table 1 on page 3.

REFERENCES

1. M. Guggenheim, *Die Biogenen Amine, und Ihre Bedeutung für die Physiologie und Pathologie des Pflanzlichen und Tierischen Stoffwechsel*, Verlag Julius Springer, Berlin (1924); reprinted by S. Karger, Basel (1951).

2. *Specifications and Criteria for Biochemical Compounds*, R. S. Tipson (Ed.), Third Edition (1972), National Academy of Sciences, Washington, D.C.

3. *United States Pharmacopeia*, XIX Revision (1975), Mack Publishing Company, Easton, Pa.

4. *The Merck Index*, Eighth Edition (1968), Merck & Co., Inc., Rahway, N.J.

5. C. B. Childs, *Low Pressure Mercury Arc for Ultraviolet Calibration*, Appl. Opt., **1**, 711–716 (1962).

6. *Manual on Recommended Practices in Spectrophotometry*, Third Edition, American Society for Testing Materials, Philadelphia, Pa. (1969).

7. J. R. Edisburg, *Practical Hints on Absorption Spectrometry*, Plenum Press, New York, N.Y. (1968).

8. W. J. Youden, *Statistical Techniques for Collaborative Tests*, Association of Official Analytical Chemists, Washington, D.C. (1967).

9. R. N. Rand, *Practical Spectrophotometric Standards*, Clin. Chem., **15**, 839–863 (1969).

10. R. E. Vanderlinde, A. H. Richards, and P. Kowalski, *Linearity and Accuracy of Ultraviolet and Visible Wavelength Photometers: An Inter-Laboratory Survey*, Clin. Chim. Acta, **61**, 39–46 (1975).

11. S. Udenfriend, *Fluorescence Assay in Biology and Medicine*, Vols. I and II, Academic Press, Inc. (1969).

12. L. T. Grady, S. E. Hays, R. H. King, H. R. Klein, W. J. Mader, D. K. Wyatt, and R. O. Zimmer, *Drug Purity Profiles*, J. Pharm. Sci., **62**, 456–464 (1973).

13. M. Knowlton, F. C. Dohan, and H. Sprince, *Use of Modified Ehrlich's Reagent for Measurement of Indolic Compounds*, Anal. Chem., **32**, 666–668 (1960).

14. H. Weil-Malherbe, *Detection of Biochemical Compounds, Amines*, Chap. 21, part 3, pp. 518–525, in *Data for Biochemical Research*, R. M. C. Dawson, D. C. Elliott, W. H. Elliott, and K. M. Jones, eds., Oxford University Press, Oxford, U.K., and New York (1969).

Acknowledgments

The technical assistance of Dr. Robert A. Anderson, Dr. Betty Lubitz, Harold Bauer, A. V. R. Crain, John Schmuck, and Dr. Robert Murphy, who measured and reviewed data, is gratefully acknowledged. A note of thanks is due to Dr. Waldo E. Cohn, Director of the NAS–NRC Office of Biochemical Nomenclature, especially for his advice regarding approved nomenclature, and also for his assistance in editing the manuscript. The Subcommittee also expresses its appreciation to Dr. Donald L. MacDonald, chairman of the parent committee, for encouragement to complete this work.

BA-1

D-(−)-Amphetamine sulfate
IUPAC: (−)-α-Methylphenethylamine sulfate (2 : 1)
C.A.: (R)-(−)-α-Methylbenzeneethanamine sulfate
    (2 : 1)
Other: (levamphetamine sulfate, *l*-amphetamine sulfate)

Formula: $(C_9H_{13}N)_2 \cdot H_2SO_4$
Formula Wt.: 368.5

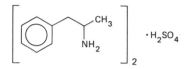

Description:

Appearance: Colorless crystalline powder.
Melting Point: >300° (dec.).
Solubility: Soluble in water.
Stability: Stable at room temperature.
Absorption Spectrum: In 0.1 M HCl, λ_{max} 207 nm, ϵ ~16,800; λ_{max} 257 nm, ϵ ~410. Free base in 0.1 M HCl has λ_{max} 207, ϵ ~8,630.
Fluorescence Spectrum: In water, λ_{ex} 255 nm, 303 nm, and 335 nm, λ_{em} 407 nm. In 0.05 M sodium acetate buffer, pH 4, λ_{ex} 259 nm, 311 nm, and 340 nm, λ_{em} 410 nm.
Anion Test: Positive test for sulfate.[1]
Homogeneity: Thin-layer Chromatography
System 1. Spot 5 μl of a 0.5% solution in methanol on silica gel F254 Merck (precoated plate). Develop with ethyl acetate : methanol : ammonia (18:1:1) to a height of 10–11 cm. Detect with UV, or with consecutive sprays of ethyl alcohol : acetic acid (1:1), pH 6 buffer, ninhydrin, then heat. R_f ~0.38.
System 2. Spot 5 μl of a 0.5% solution in methanol on silica gel F254 Merck (precoated plate). Develop with butanol : acetic acid : water (7:1:2) to a height of 10–11 cm. Detect with UV or with consecutive sprays of ethyl alcohol : acetic acid (1:1), pH 6 buffer, ninhydrin, then heat. R_f ~0.48.
Specific Rotation: $[\alpha]_D$ −22.2°, c = 4 g/100 ml, H_2O.
Source: Resolution of (±)-amphetamine.[2,3,4]
Storage: Protect from light and air.

References

1. *The U.S. Pharmacopeia*, XIX Edition (1975), p. 617.
2. O. Yu. Magidson and G. A. Garkusha, *J. Gen. Chem.* (USSR), 11, 339 (1941).
3. O. Cervinka, E. Kroupova, and O. Belovsky, *Collect. Czech. Chem. Commun.*, 33, 3551 (1968).
4. F. P. Nabenhauer, U.S. 2,276,508 (1942).

BA-2

L-(+)-Amphetamine sulfate
IUPAC: (+)-α-Methylphenethylamine sulfate (2:1)
C.A.: (S)-(+)-α-Methylbenzeneethanamine sulfate
    (2:1)
Other: (dextroamphetamine sulfate,[1] *d*-amphetamine sulfate)

Formula: $(C_9H_{13}N)_2 \cdot H_2SO_4$
Formula Wt.: 368.5

Description:

Appearance: Colorless crystalline powder.
Melting Point: >300° (dec.).
Solubility: Soluble in water, slightly soluble in alcohol.
Stability: Stable at room temperature.
Absorption Spectrum; In 0.1 M HCl, λ_{max} 206 nm, ϵ ~18,500; λ_{max} 257 nm, ϵ ~370.
Fluorescence Spectrum: In water λ_{ex} 255 nm, 311 nm, and 335 nm, λ_{em} 407 nm. In 0.05 M sodium acetate buffer, pH 4, λ_{ex} 263 nm, 305 nm, and 319 nm, λ_{em} 407 nm.
Anion Test: Positive for sulfate.[2]
Homogeneity: Thin-layer Chromatography
System 1. Spot 5 μl of a 0.5% solution in methanol on silica gel F254 Merck (precoated plate). Develop with ethyl acetate : methanol : ammonia (18:1:1) to a height of 10–11 cm. Detect with UV, or with consecutive sprays of ethyl alcohol : acetic acid (1:1), pH 6 buffer, ninhydrin, then heat. R_f ~0.38.
System 2. Spot 5 μl of a 0.5% solution in methanol on silica gel F254 Merck (precoated plate). Develop with butanol : acetic acid : water (7:1:2) to a height of 10–11 cm. Detect with UV or with ninhydrin reagent followed by gentle heating. R_f ~0.48.
Specific Rotation: $[\alpha]_D$ +21.1°, c = 4 g/100 ml, H_2O.
Source: Reductive amination of phenylacetone,[3] resolution of (±)-amphetamine with (+)-tartaric acid, followed by treatment with 10% H_2SO_4.[4-7]
Likely Impurities: (−)-Amphetamine sulfate.
Storage: Protect from light and air.

References

1. *The U.S. Pharmacopeia*, XIX Revision (1975), p. 126.
2. *The U.S. Pharmacopeia*, XIX Revision (1975), p. 617.
3. L. Haskelberg, *J. Am. Chem. Soc.*, 70, 2811 (1948).
4. F. P. Nabenhauer, U.S. 2,276,508 (1942).
5. T. H. Temmler, Brit. pat. 508,757 (1939).
6. O. Yu. Magidson and G. A. Garkusha, *J. Gen. Chem.* (USSR), 11, 339 (1941); Chem. Abstr., 35, 5868 (1941).
7. O. Cervinka, E. Kroupova, and O. Belovsky, *Collect. Czech. Chem. Commun.*, 33, 3551 (1968).

Additional References

J. C. Craig, R. P. K. Chan, and S. K. Roy, *Tetrahedron*, 23, 3573 (1967); specific rotation, optical rotary dispersion curves and configuration. V. M. Potapov. V. M. Dem'yanovich, and A. P. Terent'ev, *J. Gen. Chem.* (USSR), 33, 2311 (1963); configuration.

R. J. Warren, P. P. Begosh, and J. E. Zarembo, *J. Assoc. Offic. Anal. Chem*, 54, 1179 (1971); IR, UV and NMR spectral data.

M. Donike, *J. Chromatogr.*, 103, 91 (1975); derivatives for gas chromatography–mass spectrometry.

J. A. F. de Silva and N. Strojny, *Anal. Chem.*, 47, 714 (1975); reaction product fluorescence with fluorescamine.

J. E. Wallace, J. D. Biggs, and S. L. Ladd, *Anal. Chem.*, 40, 2207 (1968); UV absorption of cerium oxidation product.

K. F. Harbaugh, C. M. O'Donnell, and J. D. Winefordner, *Anal. Chem.*, 46, 1206 (1974); pulsed source, time-resolved phosphorimetry.

BA-3

3,4-Dimethoxyphenethylamine hydrochloride
IUPAC: 3,4-Dimethoxyphenethylamine hydrochloride
C.A.: 3,4-Dimethoxybenzeneethanamine hydrochloride
Other: (homoveratrylamine hydrochloride)

Formula: $C_{10}H_{16}ClNO_2$
Formula Wt.: 217.7

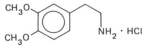

Description:

Appearance: Colorless crystals; acceptable preparations may be light tan.

Melting Point: ~152–154° (after drying over P_2O_5).

Solubility: Soluble in water and alcohol.

Stability: Normal phenethylamine characteristics.

Absorption Spectrum: In 0.1 *M* HCl, λ_{max} 228 nm, ϵ ~8,200; λ_{max} 278 nm, ϵ ~2,880.

Fluorescence Spectrum: In water, λ_{ex} 247 nm and 279 nm, λ_{em} 336 nm. In 0.05 *M* sodium acetate buffer, pH 4, λ_{ex} 279 nm, λ_{em} 333 nm.

Anion Test: Positive test for chloride.[1]

Homogeneity: Thin-layer Chromatography

System 1. Spot 8 μl of a 0.5% solution in water on silica gel F254 Merck (precoated plate). Develop with butanol : acetic acid : water (7:1:2) to a height of 10–11 cm. Detect with uv light. R_f ~0.53.

System 2. Spot 8 μl of a 0.5% solution in water on cellulose F Merck (precoated plate). Develop with butanol : acetic acid : water (7:1:2) to a height of 10–11 cm. Detect with uv light. R_f ~0.69.

Source: Synthesized by reduction of 3,4-dimethoxyphenyl-acetonitrile[2,3] or 3,4-dimethoxy-β-nitrostyrene.[4,5]

Stability: Protect from light and air.

References

1. *The U.S. Pharmacopeia*, XIX Revision (1975), p. 616.
2. G. Hahn and O. Schales, *Chem. Ber.*, **67**, 1486 (1934).
3. A. E. Bide and P. A. Wilkinson, *Chem. Ind. (London)*, **64**, 85 (1945).
4. A. Skita and F. Keil, *Chem. Ber.*, **65**, 424 (1932).
5. *Beilstein's Handbook*, **13**, 800; II 325; III 486; IV 2207.

Additional References

J. Lundstrom and S. Agurell, *J. Chromatogr.*, **36**, 165 (1968); gas chromatography.

A. A. Boulton, *Nature*, **231**, 22 (1971); a review including occurrence and identification in mammals.

R. L. Jones, P. Bory, W. T. Brown and P. L. McGeer, *Can. J. Biochem.*, **47**, 185 (1969); gas chromatography and two dimensional paper chromatography.

BA-4

N,N-Dimethyltryptamine

IUPAC: 3-[2-(Dimethylamino)ethyl]indole

C.A.: *N,N*-Dimethyl-1*H*-indole-3-ethanamine

Other: (DMT)

Formula: $C_{12}H_6N_2$

Formula Wt.: 188.3

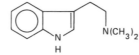

Description:

Appearance: Colorless crystals; acceptable preparations may be light pink or tan.

Melting Point: Two polymorphic forms have been reported,[1] ~47–49° and ~71–73°. Acceptable preparations may melt between these limits.

Solubility: Soluble in dilute acetic and dilute mineral acids; recrystallized from hexane.

Stability: Forms an amine oxide.[1,2]

Absorption Spectrum: In 0.1 *M* HCl, λ_{max} 218 nm, ϵ ~37,700; λ_{max} 280 nm, ϵ ~6,080.

Fluorescence Spectrum: In 0.05 *M* sodium acetate buffer at pH 4, λ_{ex} 280 nm, λ_{em} 366 nm.

Homogeneity: Thin-layer chromatography

System 1. Spot 5 μl of a 0.5% solution in methanol on silica gel F254 Merck (precoated plate). Develop with ethyl acetate : methanol : ammonia (19:1:1) to a height of 10–11 cm. Detect with iodoplatinate reagent or with uv. R_f ~0.45.

System 2. Spot 5 μl of a 0.5% solution in methanol on silica gel F254 Merck (precoated plate). Develop with ethyl acetate : methanol : isopropylamine (48:1:1) to a height of 10–11 cm. Detect with iodoplatinate reagent or with uv. R_f ~0.17.

Source: Complex metal hydride reduction of *N,N*-dimethyl-3-indoleacetamide,[2] *N,N*-dimethyl-3-indoleglyoxylamide,[3] or via an indole Grignard.[4]

Storage: Protect from light and air.

References

1. M. S. Fish, N. M. Johnson, and E. C. Horning, *J. Am. Chem. Soc.*, **77**, 5892 (1955).
2. M. S. Fish, N. M. Johnson, and E. C. Horning, *J. Am. Chem. Soc.*, **78**, 3668 (1956).
3. H. Kondo, H. Kataoka, Y. Hayashi, and T. Dodo, *Itsuu Kenkyusho Nempo*, **10**, 1 (1959); *Chem. Abstr.*, **54**, 492 (1960).
4. C. R. Ganellin and H. F. Ridley, *Chem. Ind. (London)*, **31**, 1388 (1964).

Additional References

L. A. Cohen, J. W. Daly, H. Kny, and B. Witkop, *J. Am. Chem. Soc.*, **82**, 2184 (1960); nuclear magnetic resonance spectrum.

P. Baumann and N. Narasimhachari, *J. Chromatogr.*, **86**, 269 (1973); thin-layer chromatography on cellulose.

T. R. Bosin and C. Wehler, *J. Chromatogr.*, **75**, 126 (1973); thin-layer chromatography on silica gel.

R. P. Maickel and F. P. Miller, *Anal. Chem.*, **38**, 1937 (1966); weakly fluorescent reaction product with *o*-phthalaldehyde.

BA-5

Dopamine hydrochloride

IUPAC: 4-(2-Aminoethyl)pyrocatechol hydrochloride

C.A.: 4-(2-Aminoethyl)-1,2-benzenediol hydrochloride

Other: (3,4-dihydroxyphenethylamine hydrochloride, 3-hydroxytyramine hydrochloride)

Formula: $C_8H_{12}ClNO_2$

Formula Wt.: 189.6

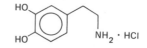

Description:

Appearance: Colorless crystals; acceptable preparations may be slightly off-white.

Melting Point: ~245° (dec.).

Solubility: Freely soluble in water, soluble in methanol and insoluble in most of the other organic solvents.

Stability: Relatively stable in water for short periods of time when protected from oxidants.

Absorption Spectrum: In 0.1 *M* HCl, λ_{max} 219 nm, ϵ ~6,490; λ_{max} 280 nm, ϵ ~2,750.

Fluorescence Spectrum: In water, λ_{ex} 283 nm, λ_{em} 336 nm. In 0.05 *M* sodium acetate buffer, λ_{ex} 283 nm, λ_{em} 336 nm.

Reaction Product Fluorescence: Sodium periodate oxidation,[1] λ_{ex} 313 nm, λ_{em} 393 nm.

Anion Test: Positive test for chloride ion.[2]

Homogeneity: Thin-layer Chromatography

System 1. Spot 4 μl of a 0.5% solution in water on silica gel F254 Merck (precoated plate). Develop with butanol : acetic acid : water (7:1:2) to a height of 10–11 cm. Detect with Folin's reagent followed by a spray of 10% sodium carbonate. R_f ~0.45.

System 2. Spot 4 μl of a 0.5% solution in water on cellulose F Merck (precoated plate). Develop with butanol : acetic acid : water (7 : 1 : 2) to a height of 10–11 cm. Detect with Folin's reagent followed by a spray of 10% sodium carbonate. R_f ~0.43.

Source: Synthesized by acidic *O*-demethylation of 3,4-dimethoxyphenethylamine,[3-5] which is prepared by reduction of 3,4-dimethoxy-β-nitrostyrene or 3,4-dimethoxyphenyl-acetonitrile.[6]

Storage: Protect from light and air.

References

1. A. Anton and D. F. Sayre, *J. Pharmacol. Exp. Ther.*, **145**, 326 (1964).
2. *The U.S. Pharmacopeia*, XIX Revision, (1975), p. 616.
3. G. Hahn and K. Stiehl, *Chem. Ber.*, **69**, 2640 (1936).
4. G. R. Clemo, F. K. Duxbury, and G. A. Swann, *J. Chem. Soc.*, 3464 (1952).
5. G. A. Swann and D. Wright, *J. Chem. Soc.*, 381 (1954); synthesized carbon-14 labeled dopamine.
6. A. E. Bide and P. A. Wilkinson, *Chem. Ind. (London)*, **64**, 84 (1945).

Additional References

S. Sinoh, C. R. Creveling, S. Udenfriend, and B. Witkop, *J. Am. Chem. Soc.*, **81**, 6236 (1959); the chemical and enzymatic oxidation of dopamine is described.

S. Sinoh and B. Witkop, *J. Am. Chem. Soc.*, **81**, 6222 and 6231 (1959); chemical, spectrophotometric, and polarographic studies on dopamine and its rearrangement products are described.

G. W. A. Milne, H. M. Fales, and R. W. Colburn, *Anal. Chem.*, **45**, 1952 (1973); chemical ionization mass spectrometry.

S. H. Koslow, F. Cattabeni, and E. Costa, *Science*, **176**, 177 (1972); determination by mass fragmentography.

System 2. Spot 4 μl of a 0.5% solution in water containing a few drops of formic acid on cellulose F Merck (precoated plate). Develop with butanol : acetic acid : water (7 : 1 : 2) to a height of 10–11 cm. Detect with uv light or with a spray of Folin's reagent followed by a spray of 10% aqueous sodium carbonate. R_f ~0.37.

Specific Rotation: $[\alpha]_D^{25}$ −50 to −53.5°, *c* = 2 g/100 ml, 0.5 *M* HCl.[1] Optical purity determined on the *N*-acetyl-3,4-di-*O*-acetyl derivative, $[\alpha]_D^{21}$ −94.7°, *c* = 1.01 g/100 ml, CHCl$_3$.[3]

Source: Resolution of (±)-epinephrine with (+)-tartaric acid.

Likely Impurities: (+)-Epinephrine.

Storage: Protect from light and air.

References

1. *The U.S. Pharmacopeia*, XIX Revision (1975), p. 169.
2. L. C. Schroeter, T. Higuchi, and E. E. Schuler, *J. Pharm. Sci.*, **47**, 723 (1958).
3. L. H. Welsh, *J. Am. Chem. Soc.*, **74**, 4967 (1952). P. Pratesi, A. La Manna, A. Campiglio, and V. Ghislandi, *J. Chem. Soc.*, 2069 (1958), report $[\alpha]_D^{20}$ −87.4°, *c* = 1 g/100 ml, CHCl$_3$ for the triacetyl derivative, and the configuration of natural (−)-epinephrine is the same as D-(−)-mandelic acid.

Additional References

J. C. Craig and S. K. Roy, *Tetrahedron*, **21**, 1847 (1965); the optical rotary dispersion curve for D-(−)-epinephrine.

P. Cancalon and J. D. Klingman, *J. Chromatogr. Sci.*, **10**, 253 (1972); gas-liquid chromatography of derivative.

J. Merzhauser, E. Roder, and Ch. Hesse, *Klin. Wochenschr.*, **51**, 883 (1973); high-pressure liquid chromatography of triacetyl derivative.

S. H. Koslow and M. Schlumpf, *Nature*, **251**, 530 (1974); assay by mass-fragmentography.

L. Pichat and M. Audinot, *Bull. Soc. Chim. Fr.*, 2255 (1961); synthesis of carbon-14 labeled (±)-epinephrine.

BA-6

D-(−)-Epinephrine

IUPAC: (−)-3,4-Dihydroxy-α-[(methylamino)methyl] benzyl alcohol

C.A.: (*R*)-(−)-4-[1-Hydroxy-2-(methylamino)ethyl]-1,2-benzenediol

Other: (epinephrine,[1] adrenaline)

Formula: $C_9H_{13}NO_3$
Formula Wt.: 183.2

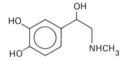

Description:

Appearance: Colorless microcrystals; acceptable preparations may be slightly off-white.

Melting Point: ~209–210° (dec.)

Solubility: In water, ~0.1 g/100 ml; insoluble in alcohol and most other organic solvents.

Stability: Sensitive to air, light, heat, and alkalis. Metals, notably copper, iron, and zinc, destroy its activity. In solution with sulfite or bisulfite it slowly forms an inactive sulfonate.[2] The red color which forms when neutral or alkaline solutions are exposed to air is caused by adrenochrome.

Absorption Spectrum: 0.1 *M* HCl, λ_{max} 221 nm, ϵ ~6,100; λ_{max} 280 nm, ϵ ~2,700.

Fluorescence Spectrum: In 0.05 *M* sodium acetate buffer, pH 4, λ_{ex} 283 nm, λ_{em} 337 nm.

Homogeneity: Thin-layer Chromatography

System 1. Spot 4 μl of a 0.5% solution in water containing a few drops of formic acid on silica gel F254 Merck (precoated plate). Develop with butanol : acetic acid : water (7 : 1 : 2) to a height of 10–11 cm. Detect with uv light or with a spray of Folin's reagent followed by a spray of 10% aqueous sodium carbonate, R_f ~0.23.

BA-7

D-(−)-Epinephrine (+)-tartrate

IUPAC: (−)-3,4-Dihydroxy-α-[(methylamino)methyl] benzyl alcohol (+)-tartrate

C.A.: (*R*)-(−)-4-[1-Hydroxy-2-(methylamino)ethyl]-1,2-benzenediol [*R*-(*R**,*R**)]-2,3-dihydroxybutane-dioate (1 : 1) (salt)

Other: (epinephrine bitartrate,[1] adrenaline bitartrate)

Formula: $C_{13}H_{19}NO_9$
Formula Wt.: 333.3

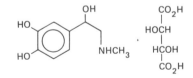

Description:

Appearance: Colorless crystals; acceptable preparations may be greyish-white.

Melting Point: ~148–152° (dec.).

Solubility: Soluble in water, slightly soluble in alcohol, and insoluble in most of the other organic solvents.

Stability: Unstable in alkaline solutions.

Absorption Spectrum:[2] In 0.1 *M* HCl, λ_{max} 220 nm, ϵ ~6,440; λ_{max} 279 nm, ϵ ~2,850.

Fluorescence Spectrum:[3] In water, λ_{ex} 287 nm, λ_{em} 337 nm. In 0.05 *M* sodium acetate buffer, λ_{ex} 282 nm, λ_{em} 337 nm.

Reaction Product Fluorescence: Iodine oxidation, λ_{ex} 421 nm, λ_{em} 514 nm.

Anion Test: Positive test for tartrate.[4]

Homogeneity: Thin-layer Chromatography

System 1. Spot 4 μl of a 0.5% solution in water on silica gel F254 Merck (precoated plate). Develop with butanol : acetic acid : water (7 : 1 : 2) to a height of 10–11 cm. Detect with UV or with a spray of Folin's reagent followed by a spray of 10% sodium carbonate. R_f ~0.23.

System 2. Spot 4 μl of a 0.5% solution in water on cellulose F Merck (precoated plate). Develop with butanol : acetic acid : water (7 : 1 : 2) to a height of 10–11 cm. Detect with UV or with a spray of Folin's reagent followed by a spray of 10% sodium carbonate. R_f ~0.31.

Specific Rotation: Optical purity determined on the *N*-acetyl-3,4-di-*O*-acetyl derivative, $[\alpha]_D^{21}$ −94.7°, c = 1.01 g/100 ml, CHCl₃.[5]

Source: Reduction of (methylamino)acetylcatechol followed by resolution of (±)-epinephrine with (+)-tartaric acid.[6,7]

Likely Impurities: (+)-Epinephrine

Storage: Protect from light and air.

References

1. *The U.S. Pharmacopeia*, XIX Revision (1975), p. 171.
2. T. Kappe and M. D. Armstrong, *J. Med. Chem.*, **8**, 368 (1965).
3. Z. Kahane and P. Vestergaard, *J. Lab. Clin. Med.*, **65**, 848 (1969).
4. *The U.S. Pharmacopeia*, XIX Revision (1975), p. 617.
5. L. H. Welsh, *J. Am. Chem. Soc.*, **74**, 4967 (1952).
6. K. R. Payne, *Ind. Chem.*, **37**, 523 (1961).
7. H. Loewe, *Arzneim.-Forsch.*, **4**, 583 (1954).

BA-8

L-(+)-Epinephrine (−)-tartrate

IUPAC: (+)-3,4-Dihydroxy-α-[(methylamino)methyl]benzylalcohol (−)-tartrate

C.A.: (*S*)-(+)-4-[1-Hydroxy-2-(methylamino)ethyl]-1,2-benzenediol [*S*-(*R**,*R**)]-2,3-dihydroxy-butanedioate (1 : 1) (salt)

Formula: $C_{13}H_{19}NO_9$
Formula Wt.: 333.3

Description:

Appearance: Colorless crystals.

Melting Point: ~147–148.5° (dec.).

Solubility: Soluble in water and insoluble in most organic solvents.

Absorption Spectrum: In 0.1 *M* HCl, λ_{max} 220 nm, ε ~6,650; λ_{max} 279 nm, ε ~2,900.

Fluorescence Spectrum: In water, λ_{ex} 287 nm, λ_{em} 337 nm. In 0.05 *M* sodium acetate buffer, λ_{ex} 283 nm, λ_{em} 337 nm.

Reaction Product Fluorescence:[1] Iodine oxidation, λ_{ex} 421 nm, λ_{em} 514 nm.

Anion Test: Positive test for tartrate.[2]

Homogeneity: Thin-layer Chromatography

System 1. Spot 4 μl of a 0.5% solution in water on silica gel F254 Merck (precoated plate). Develop with butanol : acetic acid : water (7 : 1 : 2) to a height of 10–11 cm. Detect with UV or with a spray of Folin's reagent followed by a spray of 10% sodium carbonate. R_f ~0.23.

Specific Rotation: Optical purity can be determined on the *N*-acetyl-3,4-di-*O*-acetyl derivative.[3]

Source: Reduction of (methylamino)acetylcatechol to (±)-epinephrine, followed by resolution. This "dextro" form of epinephrine does not occur naturally in animals and humans.

Likely Impurities: (−)-Epinephrine.

Storage: Protect from light and air.

References

1. Z. Kahane and P. Vestergaard, *J. Lab. Clin. Med.*, **65**, 848 (1965).
2. *The U.S. Pharmacopeia*, XIX Revision (1975), p. 617.
3. L. H. Welsh, *J. Am. Chem. Soc.*, **74**, 4967 (1952).

BA-9

Gramine

IUPAC: 3-[(Dimethylamino)methyl]indole

C.A.: *N,N*-Dimethyl-1*H*-indole-3-methanamine

Other: (Donaxine)

Formula: $C_{11}H_{14}N_2$
Formula Wt.: 174.25

Description:

Appearance: Colorless needles or plates.

Melting Point: ~132–133°.

Solubility: Soluble in alcohol; practically insoluble in water; recrystallized from acetone.[1]

Stability: Susceptible to oxidation at 2-indolyl position, and forms amine oxide.[2]

Absorption Spectrum: In 0.1 *M*, HCl, λ_{max} 214 nm, ε ~46,000; λ_{max} 269 nm, ε ~7,060; λ_{max} 278 nm, ε ~6,960; λ_{max} 286 nm, ε ~5,550.

Fluorescence Spectrum: In 0.05 *M* sodium acetate buffer, pH 4, λ_{ex} 280 nm, λ_{em} 348 nm.

Homogeneity: Thin-layer Chromatography

System 1. Spot 5 μl of a 0.5% solution in methanol on silica gel F254 Merck (precoated plate). Develop with ethyl acetate : methanol : ammonia (18 : 1 : 1) to a height of 10–11 cm. Detect with UV or with iodoplatinate reagent. R_f ~0.33.

System 2. Spot 5 μl of a 0.5% solution in methanol on silica gel F254 Merck (precoated plate). Develop with ethyl acetate : methanol : isopropylamine (48 : 1 : 1) to a height of 10–11 cm. Detect with UV or with iodoplatinate reagent R_f ~0.14.

Source: Synthesized from indole.[1,3]

Likely Impurities: Indole

Storage: Protect from air.

References

1. T. Nogradi, *Monatsh. Chem.*, **88**, 768 (1957).
2. D. W. Henry and E. Leete, *J. Am. Chem. Soc.*, **79**, 5254 (1957).
3. H. R. Snyder, C. W. Smith, and J. M. Stewart, *J. Am. Chem. Soc.*, **66**, 200 (1944).

Additional Reference

Y. Kanaoka, Y. Ban, T. Oishi, O. Yonemitsu, M. Terashima, T. Kimura, and M. Nakagawa, *Chem. Pharm. Bull.*, **8**, 294 (1960); infrared spectrum.

BA-10
Histamine dihydrochloride

IUPAC: Histamine dihydrochloride or 4-(2-aminoethyl)imidazole dihydrochloride

C.A.: 1H-Imidazole-4-ethanamine dihydrochloride

Other: (4-imidazoleethylamine hydrochloride, β-aminoethylimidazole hydrochloride, 2-[4-imidazole]ethylamine hydrochloride)

Formula: $C_5H_{11}Cl_2N_3$
Formula Wt.: 184.1

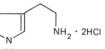

Description:

Appearance: Colorless prisms.

Melting Point: ~247–249°.

Solubility: Freely soluble in water, methanol; soluble in ethanol.

Stability: Sensitive to light.

Absorption Spectrum: In 0.1 M HCl, λ_{max} 211 nm, ϵ ~5,640.

Fluorescence Spectrum: In water, λ_{ex} 259 nm and 315 nm, λ_{em} 410 nm. In 0.05 M sodium acetate buffer, pH 4, λ_{ex} 262 nm, 325 nm and 335 nm, λ_{em} 410 nm.

Reaction Product Fluorescence: Condensation with o-phthalaldehyde,[1] at 0.02 to 0.1 μg/ml, λ_{ex} 339 nm, λ_{em} 450 nm.

Anion Test: Positive test for chloride.[2]

Homogeneity: Thin-layer Chromatography

System 1. Spot 4 μl of a 0.5% solution in water on silica gel F254 Merck (precoated plate). Develop with butanol : acetic acid : water (7:1:2) to a height of 10–11 cm. Detect with a spray of ninhydrin reagent followed by gentle heating. R_f ~0.06.

System 2. Spot 4 μl of a 0.5% solution in water on cellulose F Merck (precoated plate). Develop with butanol : acetic acid : water (7:1:2) to a height of 10–11 cm. Detect with a spray of ninhydrin reagent followed by gentle heating. R_f ~0.15.

Source: Synthetic.[3–5]

Storage: Protect from light and air.

References

1. D. von Redlich and D. Glick, *Anal. Biochem.*, **10**, 459 (1965); reaction method described.
2. *The U.S. Pharmacopeia*, XIX Revision (1975), p. 616.
3. K. K. Koessler and M. T. Hanke, *J. Am. Chem. Soc.*, **40**, 1716 (1918).
4. B. Garforth and F. L. Pyman, *J. Chem. Soc.*, 489 (1935).
5. S. Akabori and S. Numano, *Bull. Chem. Soc. Japan*, **11**, 214 (1936).

Additional References

J. A. Oates, E. Marsh, and A. Sjoerdsma, *Clin. Chim. Acta*, **7**, 488 (1962); o-phthalaldehyde reaction product fluorescence for histamine in urine.

W. Lorenz, H. J. Reimann, H. Barth, J. Kusche, R. Meyer, A. Doenicke, and M. Hutzel, *Hoppe-Seylers Z. Physiol. Chem.*, **353**, 911 (1972); o-phthalaldehyde reaction product fluorescence for histamine in blood.

W. Lorenz, H. Barth, H. E. Kargas, A. Schmal, P. Dormann, and I. Niemeyer, *Agent Action*, **4/5**, 324 (1974); thin layer chromatography.

T. L. Perry and W. A. Schroeder, *J. Chromatogr.*, **12**, 358 (1963); paper chromatography.

P. Cancalon and J. D. Klingman, *J. Chromatogr. Sci.*, **10**, 253 (1972); gas-liquid chromatography.

G. W. A. Milne, H. M. Fales and R. W. Coburn, *Anal. Chem.*, **45**, 1952 (1973); chemical ionization mass spectrometry.

P. A. Shore, A. Burkhalter, and V. H. Cohn, Jr., *J. Pharmacol. Exp. Ther.*, **127**, 182 (1959); reaction product fluorescence with o-phthalaldehyde.

J. W. Noah and A. Brand, *J. Lab. Clin. Med.*, **62**, 506 (1963); reaction product fluorescence with benzaldehyde.

C. F. Code and F. C. McIntire, in *Methods of Biochemical Analysis*, D. Glick, ed., **3**, 49–99 (1956), Interscience–Wiley, New York; methods for purification, detection and quantitative determination of histamine.

BA-11
Hordenine sulfate dihydrate

IUPAC: p-[(2-Dimethylamino)ethyl]phenol sulfate dihydrate (2:1:2) (salt)

C.A.: 4-[(2-Dimethylamino)ethyl]phenol sulfate dihydrate (2:1:2) (salt)

Other: (N,N-dimethyltyramine sulfate dihydrate, anhaline sulfate dihydrate)

Formula: $(C_{10}H_{15}NO)_2 \cdot H_2SO_4 \cdot 2H_2O$
Formula Wt.: 464.5

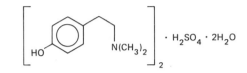

Water Content: 7.75%

Description:

Appearance: Small colorless prisms.

Melting Point: Loses water >100°, resolidifies then melts at ~210–212°.

Solubility: Soluble in water, slightly soluble in alcohol.

Stability: Relatively stable in aqueous solutions under anaerobic conditions.

Absorption Spectrum: In 0.1 N HCl, λ_{max} 222 nm, ϵ ~16,300; λ_{max} 275 nm, ϵ ~2,930; λ_{sh} 280 nm.[1]

Fluorescence Spectrum: In water, λ_{ex} 279 nm, λ_{em} 330 nm. In 0.05 M sodium acetate buffer, pH 4.0, λ_{ex} 282 nm, λ_{em} 333 nm.

Reaction Product Fluorescence: Formation of fluorescent derivative with nitrosonapthol,[2] at 0.02–0.2 μg/ml, λ_{ex} 465 nm, λ_{em} 553 nm.

Anion Test: Positive test for sulfate.[3]

Homogeneity: Thin-layer Chromatography

System 1. Spot 4 μl of a 0.5% solution in water on silica gel F254 Merck (precoated plate). Develop with butanol : acetic acid : water (7:1:2) to a height of 10–11 cm. Detect by spraying with Folin's reagent followed by 10% sodium carbonate. R_f ~0.35.

System 2. Spot 4 μl of a 0.5% solution in water on cellulose F Merck (precoated plate). Develop with butanol : acetic acid : water (7:1:2) to a height of 10–11 cm. Detect by spraying with Folin's reagent followed by 10% sodium carbonate. R_f ~0.67.

Source: Synthesis.[4]

Storage: Protect from light and air.

References

1. T. Kappe and M. D. Armstrong, *J. Med. Chem.*, **8**, 368 (1965); for shoulder absorption.
2. S. Spector, K. Melmon, W. Lovenborg, and A. Sjoerdsma, *J. Pharmacol. Exp. Ther.*, **140**, 229 (1963); for method used.
3. *The U.S. Pharmacopeia*, XIX Revision (1975), p. 617.
4. C.-S. Cheng, C. Ferber, R. I. Bashford, Jr., and G. F. Grillot, *J. Am. Chem. Soc.*, **73**, 4081 (1951).

Additional References

J. Lundstrom and S. Agurell, *J. Chromatogr.*, **36**, 105 (1968); gas chromatography.

J. Lundstrom and S. Agurell, *J. Chromatogr.*, **30**, 271 (1967); thin layer chromatography on silica gel G in three solvent systems.

BA-12
Melatonin
IUPAC: *N*-[2-(5-Methoxyindol-3-yl)ethyl]acetamide
C.A.: *N*-[2-(5-Methoxy-1H-indol-3-yl)ethyl]acetamide
Other: (*N*-acetyl-5-methoxytryptamine)

Formula: $C_{13}H_{16}N_2O_2$
Formula Wt.: 232.3

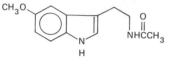

Description:

Appearance: Colorless crystals; acceptable preparations may be pale yellow.
Melting Point: ~117–119.5°.
Solubility: Recrystallized from benzene.
Stability: Susceptible to oxidation at the 2-indolyl position.
Absorption Spectrum: In 95% ethanol, λ_{max} 223 nm, ϵ ~27,500; λ_{max} 278 nm, ϵ ~6,300.[1]
Fluorescence Spectrum: In 0.05 *M* sodium acetate buffer, pH 4.0, λ_{ex} 291 nm, λ_{em} 369 nm.
Reaction Product Fluorescence: Reaction with o-phthalaldehyde,[2] at 0.1 to 0.3 μg/1.0 ml, λ_{ex} 339 nm, λ_{em} 463 nm.
Homogeneity: Thin-layer Chromatography

System 1. Spot 8 μl of a 0.5% solution in methanol–water on silica gel F254 Merck (precoated plate). Develop with butanol : acetic acid : water (7:1:2) to a height of 10–11 cm. Detect with UV, or with iodoplatinate reagent followed by a spray with 20% sulfuric acid. R_f ~0.75.

System 2. Spot 8 μl of a 0.5% solution in methanol–water on cellulose F Merck (precoated plate). Develop with butanol : acetic acid : water (7:1:2) to a height of 10–11 cm. Detect with UV, or with iodoplatinate reagent followed by a spray with 20% sulfuric acid. R_f ~0.94.
Source: Acetylation of 5-methoxytryptamine.[1,3,4]
Storage: Protect from light and air.

References
1. J. Szmuszkovicz, W. C. Anthony, and R. V. Heinzelman, *J. Org. Chem.*, **25**, 857 (1960).
2. R. P. Maickel and F. P. Miller, *Anal. Chem.*, **38**, 1937 (1966).
3. A. B. Lerner, J. D. Case, and R. V. Heinzelman, *J. Am. Chem. Soc.*, **81**, 6084 (1959).
4. J. Supniewski and S. Misztal, *Bull. Acad. Polon.*, **8**, 479 (1960).

Additional Reference

E. Cattabeni, S. H. Koslow, and E. Costa, *Science*, **178**, 166 (1972); gas chromatographic–mass spectrometric assay.

BA-13
Mescaline hydrochloride
IUPAC: 3,4,5-Trimethoxyphenethylamine hydrochloride
C.A.: 3,4,5-Trimethoxybenzeneethaneamine hydrochloride

Formula: $C_{11}H_{18}ClNO_3$
Formula Wt.: 247.7

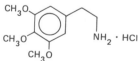

Description:

Appearance: Colorless needles.
Melting Point: ~183–184°.

Solubility: Soluble in water, alcohol.
Stability: The free base takes up CO_2 from the air to form a carbonate.
Absorption Spectrum: In 0.1 *M* HCl, λ_{max} 269 nm, ϵ ~880.
Fluorescence Spectrum: In water, λ_{ex} 259 nm, 307 nm, 335 nm, λ_{em} 407 nm. In 0.05 *M* sodium acetate buffer, pH 4.0, λ_{ex} 259 nm, λ_{em} 408 nm.
Anion Test: Positive test for chloride.[1]
Homogeneity: Thin-layer Chromatography

System 1. Spot 5 μl of a 0.5% solution in methanol on silica gel F254 Merck (precoated plate). Develop with ethyl acetate : methanol : ammonia (18:1:1) to a height of 10–11 cm. Detect with UV or with a spray of iodoplatinate reagent. R_f ~0.16.

System 2. Spot 5 μl of a 0.5% solution in methanol on silica gel F254 Merck (precoated plate). Develop with ethyl acetate : methanol : isopropylamine (48:1:1). Detect with UV or with a spray of iodoplatinate reagent. R_f ~0.07.
Source: Chemical or catalytic reduction of 3,4,5-trimethoxynitrostyrene,[2,3] 3,4,5-trimethoxybenzyl cyanide.[4,5]
Storage: Protect from air.

References

1. *The U.S. Pharmacopeia*, XIX Revision (1975), p. 616.
2. G. Hahn and F. Rumpf, *Chem. Ber.*, **71**, 2141 (1938).
3. F. Benington and R. D. Morin, *J. Am. Chem. Soc.*, **73**, 1353 (1951).
4. M. U. Tsao, *J. Am. Chem. Soc.*, **73**, 5495 (1951).
5. A. Dornow and G. Petsch, *Arch. Pharm.*, **285**, 323 (1952),

Additional References

J. Lundstrom and S. Agurell, *J. Chromatogr.*, **36**, 105 (1968); gas chromatography.
L. H. Briggs, L. D. Colebrook, H. M. Fales, and W. C. Willman, *Anal. Chem.*, **29**, 904 (1957); infrared spectrum.
E. G. C. Clark, *J. Pharm. Pharmacol.*, **9**, 187 (1957); microchemical identification.
W. Block and K. Block, *Chem. Ber.*, **85**, 1009 (1952); synthesis of carbon-14 labeled product.
J. Lundstrom and S. Agurell, *J. Chromatogr.*, **30**, 271 (1967); thin layer chromatography on silica gel G in three solvent systems.

BA-14
DL-Metanephrine hydrochloride
IUPAC: (±)-α-[(Methylamino)methyl]vanillyl alcohol hydrochloride
C.A.: (±)-4-Hydroxy-3-methoxy-α-[(methylamino)methyl]benzenemethanol hydrochloride
Other: (3-*O*-methylepinephrine hydrochloride, 3-*O*-methyladrenaline hydrochloride

Formula: $C_{10}H_{16}ClNO_3$
Formula Wt.: 233.7

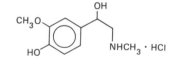

Description:

Appearance: Small colorless prisms.
Melting Point: ~170–171° (dec.).
Solubility: Water, ~20 g/100 ml.
Stability: Unstable to light and air.
Absorption Spectrum: In 0.1 *M* HCl, λ_{max} 227 nm, ϵ ~7,080; λ_{max} 278 nm, ϵ ~2,830; λ_{sh} 283 nm.[1]
Fluorescence Spectrum: In water, λ_{ex} 283 nm, λ_{em} 337 nm. In 0.05 *M* sodium acetate, pH 4.0, λ_{ex} 283 nm, λ_{em} 333 nm.

Reaction Product Fluorescence: Oxidation with periodate at pH 6.5 to form indole derivative,[3] at 0.02 to 0.1 μg/ml, λ_{ex} 421 nm, λ_{em} 514 nm.

Anion Test: Positive test for chloride.[3]

Homogeneity: Thin-layer Chromatography

System 1. Spot 4 μl of a 0.5% solution in water on silica gel F254 Merck (precoated plate). Develop with butanol : acetic acid : water (7:1:2) to a height of 10–11 cm. Detect with Folin's reagent followed by a spray with 10% sodium carbonate. R_f ~0.38.

System 2. Spot 4 μl of a 0.5% solution in water on cellulose F Merck (precoated plate). Develop with butanol : acetic acid : water (7:1:2) to a height of 10–11 cm. Detect with Folin's reagent followed by a spray with 10% sodium carbonate. R_f ~53.

Source: Synthesized by catalytic hydrogenation of α-(N-methyl-N-benzylamino)-4-hydroxy-3-ω-methylamino-4-hydroxy-3-methoxy-acetophenone,[4] and other methods.[5,6]

Storage: Protect from light and air.

References

1. T. Kappe and M. D. Armstrong, *J. Med. Chem.*, **8**, 368 (1965).
2. Z. Kahane and P. Vestergaard, *J. Lab. Clin. Med.*, **70**, 333 (1967).
3. *The U.S. Pharmacopeia*, XIX Revision (1975), p. 616.
4. F. Kulz and C. A. Hornung, Ger. pat. 682,394 (1939); *Chem. Abstr.*, **36**, 3011 (1942).
5. R. A. Heacock and O. Hutzinger, *Chem. Ind. (London)*, 595 (1961).
6. J. Axelrod, S. Senoh, and B. Witkop, *J. Biol. Chem.*, **233**, 697 (1958).

BA-15
5-Methoxytryptamine
IUPAC: 3-(2-Aminoethyl)-5-methoxyindole
C.A.: 5-Methoxy-1H-indole-3-ethanamine

Formula: $C_{11}H_{14}N_2O$
Formula Wt.: 190.2

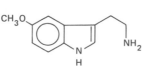

Description:

Appearance: Colorless crystals; acceptable preparations may be pale yellow or beige.

Melting Point: ~121.5–122.5°.

Solubility: Soluble in chloroform, insoluble in water, recrystallized from ethanol or benzene.

Stability: Susceptible to oxidation at the 2-indolyl position and amine group.

Absorption Spectrum: In 0.1 M HCl, λ_{max} 220 nm. ϵ ~27,700; λ_{max} 277 nm, ϵ ~7,000; λ_{sh} 296 nm and 308 nm.[1]

Fluorescence Spectrum: In 0.05 M sodium acetate buffer at pH 4.0, λ_{ex} 291 nm, λ_{em} 350 nm.

Reaction Product Fluorescence: Reaction with o-phthaldehyde,[2] at 0.1 to 0.5 μg/ml, λ_{ex} 339 nm, λ_{em} 463 nm.

Homogeneity: Thin-layer Chromatography

System 1. Spot 4 μl of a 0.5% solution in methanol–water on silica gel F254 Merck (precoated plate). Develop with butanol : acetic acid : water (7:1:2) to a height of 10–11 cm. Detect with ninhydrin reagent followed by gentle heating. R_f ~0.55.

System 2. Spot 4 μl of a 0.5% solution in methanol–water on cellulose F Merck (precoated plate). Develop with butanol : acetic acid : water (7:1:2) to a height of 10–11 cm. Detect with ninhydrin reagent followed by gentle heating. R_f ~0.60.

Source: Synthesized by reduction of 5-methoxyindole-3-acetonitrile,[1] 5-methoxyindole-3-glyoxylamide,[3,4] 5-methoxy-3-(2-nitrovinyl) indole,[1] and other methods.[5-8]

Storage: Protect from light and air.

References

1. J. Szmuszkovicz, W. C. Anthony, and R. V. Heinzelman, *J. Org. Chem.*, **25**, 857 (1960); ultraviolet and infrared spectra.
2. R. P. Maickel and F. P. Miller, *Anal. Chem.*, **38**, 1937 (1966).
3. J. Supniewski and S. Misztal, *Bull. Acad. Polon.*, **8**, 479 (1960).
4. J. Supniewski, S. Misztal, and T. Marczynski, *Diss. Pharm.*, **13**, 205 (1961).
5. E. Spath and E. Lederer, *Chem. Ber.*, **63**, 2102 (1930).
6. G. Bertaccini and T. Vitale, *Il Farmaco, Ed. Sci.*, **22**, 229 (1967).
7. R. A. Abramovitch and D. Shapiro, *Chem. Ind. (London)*, 1255 (1955).
8. M. Julia and P. Manoury, *Bull. Soc. Chim. Fr.*, 1411, and *ibid.* with C. Voillaume, 1417 (1965).

Additional References

T. R. Bosin and C. Wehler, *J. Chromatogr.*, **75**, 126 (1973); thin-layer chromatography.

F. Cattabeni, S. H. Koslow and E. Costa, *Science*, **178**, 166 (1972); gas chromatography–mass spectrometric assay.

P. Cancalon and J. D. Klingman, *J. Chromator. Sci.*, **10**, 253 (1972); derivatization for gas chromotography.

BA-16
D-(−)-Norepinephrine
IUPAC: (−)-α-(Aminomethyl)-3,4-dihydroxybenzyl alcohol
C.A.: (R)-(−)-4-(2-Amino-1-hydroxyethyl)-1,2-benzenediol
Other: (levarterenol, D-arterenol, D-noradrenaline)

Formula: $C_8H_{11}NO_3$
Formula Wt.: 169.2

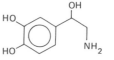

Description:

Appearance: Colorless microcrystals.

Melting Point: ~212° (dec.).

Solubility: Slightly soluble in water.

Stability: The compound is unstable in light and air, especially at neutral and alkaline pH. Oxidation to noradrenochrome occurs in the presence of oxygen and such divalent metal ions as copper, manganese and nickel.

Absorption Spectrum: In 0.1 M HCl, λ_{max} 221 nm, ϵ ~5,860; λ_{max} 279 nm, ϵ ~2,560.

Fluorescence Spectrum: In 0.05 M sodium acetate buffer, pH 4, λ_{ex} 283 nm, λ_{em} 337 nm.

Reaction Product Fluorescence: Iodine oxidation in alkaline ascorbate,[1] λ_{ex} 412 nm, λ_{em} 505 nm.

Other Tests: Color reactions with various reagents.[2]

Homogeneity: Thin-layer Chromatography

System 1. Spot 4 μl of a 0.5% solution in water with a few drops of formic acid on silica gel F254 Merck (precoated plate). Develop with butanol : acetic acid : water (7:1:2) to a height of 10–11 cm. Detect with UV or Folin's reagent followed by a spray of 10% sodium carbonate. R_f ~0.27.

System 2. Spot 4 μl of a 0.5% solution in water with a few drops of formic acid on cellulose F Merck (precoated plate). Develop with butanol : acetic acid : water (7:1:2) to a height of 10–11 cm. Detect with UV or Folin's reagent. R_f ~0.32.

Specific Rotation: $[\alpha]_D^{25}$ −37.3°, c = 5 g/100 ml water containing 1 equiv. hydrochloric acid.[3]

Optical purity determined on the *N*-acetyl-3,4-di-*O*-acetyl derivative, $[\alpha]_D^{25}$ −81.3°, c = 1 g/100 ml, $CHCl_3$.[4]

Source: Resolution of (±)-norepinephrine.[3]

Likely Impurities: (+)-Norepinephrine.

Storage: Protect from light and air.

References

1. P. A. Shore and J. S. Olin, *J. Pharmacol. Exp. Ther.*, **122**, 295 (1958).
2. C. F. Schwender, *Analytical Profiles of Drug Substances*, K. Florey (ed.), Academic Press, New York, Vol. 1, (1972), p. 149.
3. B. F. Tullar, *J. Am. Chem. Soc.*, **76**, 2067 (1948); U.S. Pat. 2,774,789 (Dec. 18, 1956).
4. L. H. Welsch, *J. Am. Pharm. Assoc., Sci. Ed.*, **44**, 507 (1955).

Additional References

J. Merhauser, E. Roder, and Ch. Hesse, *Klin. Wochenschr.*, **51**, 883 (1973); high-pressure liquid chromatography of triacetyl derivative.

S. H. Koslow, F. Cattabeni, and E. Costa, *Science*, **176**, 177 (1972); assay by mass fragmentography.

G. W. A. Milne, H. M. Fales, and R. W. Colburn, *Anal. Chem.*, **45**, 1952 (1973); chemical ionization mass spectrometry.

J. C. Craig and S. K. Roy, *Tetrahedron*, **21**, 1847 (1965); report optical rotary dispersion curve for D-(−)-norepinephrine and $[\alpha]_D^{25}$ −40°, c = 5 g/100 ml, 0.1 M HCl.

P. Pratesi, A. La Manna, A. Campiglio, and V. Ghislandi, *J. Chem. Soc.*, 4062 (1959); the configuration of natural (−)-norepinephrine is the same as D-(−)-mandelic acid.

tic acid : water (7:1:2) to a height of 10–11 cm. Detect with Folin's reagent followed by a spray of 10% sodium carbonate. R_f ~0.36.

Source: Catalytic hydrogenation of arterenone,[2-5] which is a possible impurity.

Storage: Protect from light and air.

References

1. T. Kappe and M. D. Armstrong, *J. Med. Chem.*, **8**, 368 (1965); UV absorption spectra and apparent acidic dissociation constants of some phenolic amines.
2. K. R. Payne, *Ind. Chem.*, **37**, 523 (1961).
3. D. R. Howton, J. F. Mead, and W. G. Clark, *J. Am. Chem. Soc.*, **77**, 2896 (1955).
4. H. Loewe, *Arzneim.-Forsch.*, **4**, (1954): review of syntheses.
5. W. Langenbeck and F. Fischer, *Pharmazie*, **5**, 56 (1950).

Additional References

R. W. Schayer, *J. Am. Chem. Soc.*, **75**, 1757 (1953); synthesis of alpha carbon-14 labeled compound.

Cancalon and J. D. Klingman, *J. Chromatogr. Sci.*, **10**, 253 (1972); derivatization for gas-liquid chromatography.

G. W. A. Milne, H. M. Fales, and R. W. Coburn, *Anal. Chem.*, **45**, 1952 (1973); chemical ionization mass spectrometry.

S. H. Koslow, F. Cattabeni, and E. Costa, *Science*, **176**, 177 (1972); assay by mass fragmentography.

BA-17

DL-Norepinephrine

IUPAC: (±)-α-(Aminomethyl)-3,4-dihydroxybenzyl alcohol

C.A.: (±)-4-(2-Amino-1-hydroxyethyl-1,2-benzenediol

Other: (DL-arterenol, DL-noradrenaline)

Formula: $C_8H_{11}NO_3$
Formula Wt.: 169.2

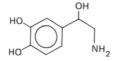

Description:

Appearance: Colorless crystals: acceptable preparations may be buff-colored.

Melting Point: ~190–191° (dec.).

Solubility: Sparingly soluble in water; slightly soluble in alcohol.

Stability: Unstable in light and air, especially at neutral and alkaline pH. In solution, it is degraded by such metals as copper, iron, or zinc, and with sulfite or bisulfite it slowly forms a sulfonate.

Absorption Spectrum: In 0.1 M HCl, λ_{max} 221 nm, ϵ ~5,820; λ_{max} 279 nm, ϵ ~2,410; λ_{sh} 285 nm.[1]

Fluorescence Spectrum: In 0.05 M sodium acetate buffer, λ_{ex} 280 nm, λ_{em} 337 nm.

Homogeneity: Thin-layer Chromatography

System 1. Spot 4 μl, of a 0.5% solution in water–formic acid on silica gel F254 Merck (precoated plate). Develop with butanol : acetic acid : water (7:1:2) to a height of 10–11 cm. Detect with Folin's reagent followed by a spray of 10% sodium carbonate, R_f ~0.39.

System 2. Spot 4 μl of a 0.5% solution in water–formic acid on cellulose F Merck (precoated plate). Develop with butanol : ace-

BA-18

DL-Norepinephrine hydrochloride

IUPAC: (±)-α-(Aminomethyl)-3,4-dihydroxybenzyl alcohol hydrochloride

C.A.: (±)-4-(2-Amino-1-hydroxyethyl)-1,2-benzenediol hydrochloride

Other: (DL-noradrenaline hydrochloride, DL-arterenol hydrochloride)

Formula: $C_8H_{12}ClNO_3$
Formula Wt.: 205.6

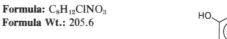

Description:

Appearance: Colorless microcrystals; acceptable preparations may be buff-colored.

Melting Point: ~143.5–145° (dec.).

Solubility: Soluble in water.

Stability: Unstable in light and air, especially at neutral and alkaline pH.

Absorption Spectrum: In 0.1 M HCl, λ_{max} 221 nm. ϵ ~6,820; λ_{max} 279 nm. ϵ ~2,910.

Fluorescence Spectrum: In water, λ_{ex} 287 nm, λ_{em} 337 nm. In 0.05 M sodium acetate buffer, λ_{ex} 283 nm, λ_{em} 337 nm.

Reaction Product Fluorescence: Iodine oxidation in alkaline L-ascorbate,[1] λ_{ex} 412 nm, λ_{em} 510 nm.

Anion Test: Positive test for chloride.[2]

Homogeneity: Thin-layer Chromatography

System 1. Spot 4 μl of a 0.5% solution in water on silica gel F254 Merck (precoated plate). Develop with butanol : acetic acid : water (7:1:2) to a height of 10–11 cm. Detect with Folin's reagent followed by a spray of 10% sodium carbonate. R_f ~0.39.

System 2. Spot 4 μl of a 0.5% solution in water on cellulose F Merck (precoated plate). Develop with butanol : acetic acid : water (7:1:2) to a height of 10–11 cm. Detect with Folin's reagent followed by a spray of 10% sodium carbonate. R_f ~0.30.

Source: Catalytic reduction of arterenone.[3]
Storage: Protect from light and air.

References

1. P. A. Shore and J. S. Olin, *J. Pharmacol. Exp. Ther.*, **122**, 295 (1958).
2. *The U.S. Pharmacopeia*, XIX Revision (1975), p. 616.
3. K. R. Payne, *Ind. Chem.*, **37**, 523 (1961).

References

1. *The U.S. Pharmacopeia*, XIX Revision (1975), p. 279.
2. C. F. Schwender, *Analytical Profiles of Drug Substances*, K. Florey (ed.) Academic Press, New York, Vol. 1, p. 149.
3. P. A. Shore and J. S. Olin, *J. Pharmacol. Exp. Ther.*, **122**, 295 (1958).
4. H. Weil-Malherbe, *Methods Biochem. Anal.*, **16**, 293 (1968).
5. *The U.S. Pharmacopeia*, XIX Revision (1975), p. 617.
6. B. F. Tullar, *J. Am. Chem. Soc.*, **70**, 2067 (1948).
7. L. H. Welsh, *J. Am. Pharm. Assoc., Sci. Ed.*, **44**, 507 (1955).

BA-19

D-(−)-Norepinephrine (+)-tartrate monohydrate
IUPAC: (−)-α-(Aminomethyl)-3,4-dihydroxybenzyl
    alcohol (+)-tartrate monohydrate
C.A.: (R)-(−)-4-(2-Amino-1-hydroxyethyl)-1,2-benzene-
    diol[R-(R*,R*)]-2,3-dihydroxybutanedioate (1 : 1)
    (salt) monohydrate
Other: (levarterenol bitartrate,[1,2] noradrenaline
    tartrate)

Formula: $C_{12}H_{19}NO_{10}$
Formula Wt.: 337.3

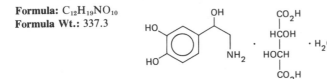

Water Content: 5.34%
Description:
  Appearance: Colorless crystals, acceptable preparations may be faintly grey.
  Melting Point: ~100–102° (turbid).
  Solubility: Soluble in water, slightly soluble in alcohol, insoluble in most of the other organic solvents.
  Stability: Unstable in light and air, especially at neutral and alkaline pH.
  Absorption Spectrum: In 0.1 M HCl, λ_{max} 222 nm. ϵ ~5,960; λ_{max} 279 nm. ϵ ~2,650.
  Fluorescence Spectrum: In water at 1.64 µg/ml, λ_{ex} 283 nm, λ_{em} 337 nm. In 0.05 M sodium acetate buffer at 1.64 µg/ml, λ_{ex} 283 nm, λ_{em} 337 nm.
  Reaction Product Fluorescence: Derived trihydroxyindole in alkaline L-ascorbate,[3,4] λ_{ex} 412 nm, λ_{em} 505 nm.
  Anion Test: Positive test for tartrate.[5]
Homogeneity: Thin-layer Chromatography
  System 1. Spot 4 µl of a 0.5% solution in water on silica gel F254 Merck (precoated plate). Develop with butanol : acetic acid : water (7 : 1 : 2) to a height of 10–11 cm. Detect with UV or with Folin's reagent followed by a spray of 10% sodium carbonate. R_f ~0.27.
  System 2. Spot 4 µl of a 0.5% solution in water on cellulose F Merck (precoated plate). Develop with butanol : acetic acid : water (7 : 1 : 2) to a height of 10–11 cm. Detect with UV or with Folin's reagent. R_f ~0.32.
Specific Rotation: $[\alpha]_D^{25}$ −10 to −12°, c = 5 g/100 ml, H_2O.[1,6] Optical purity determined on the N-acetyl-3,4-di-O-acetyl derivative, $[\alpha]_D^{25}$ −81.3°, c = 1 g/100 ml, $CHCl_3$.[7]
Source: Resolution of (±)-norepinephrine with (+)-tartaric acid.[6]
Likely Impurities: (+)-Norepinephrine.
Storage: Protect from light and air.

BA-20

DL-Normetanephrine hydrochloride
IUPAC: (±)-α-(Aminomethyl)vanillyl alcohol hydro-
    chloride
C.A.: (±)-α-(Aminomethyl)-4-hydroxy-3-
    methoxybenzenemethanol hydrochloride
Other: (3-O-methylnorepinephrine hydrochloride, 3-O-
    methylnoradrenaline hydrochloride)

Formula: $C_9H_{14}ClNO_3$
Formula Wt.: 219.7

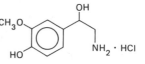

Description:
  Appearance: Small colorless prisms.
  Melting Point: ~199–200° (dec.).
  Solubility: Soluble in water >20 g/100 ml; recrystallized from absolute ethanol.
  Stability: Unstable to light and air.
  Absorption Spectrum: In 0.1 M HCl, λ_{max} 227 nm, ϵ ~7,180; λ_{max} 278 nm, ϵ ~2,830; λ_{sh} 283 nm.[1]
  Fluorescence Spectrum: In water, λ_{ex} 287 nm, λ_{em} 337 nm. In 0.05 M sodium acetate buffer, pH 4.0, λ_{ex} 282 nm, λ_{em} 333 nm.
  Reaction Product Fluorescence: Oxidation with periodate at pH 6.5 to form indole derivative,[2] at 0.02 to 0.1 µg/ml, λ_{ex} 412 nm, λ_{em} 505 nm.
  Anion Test: Positive test for chloride.[3]
Homogeneity: Thin-layer Chromatography
  System 1. Spot 4 µl of a 0.5% solution in water on silica gel F254 Merck (precoated plate). Develop with butanol : acetic acid : water (7 : 1 : 2) to a height of 10–11 cm. Detect with Folin's reagent followed by a spray with 10% sodium carbonate. R_f ~0.45.
  System 2. Spot 4 µl of a 0.5% solution in water on cellulose F Merck (precoated plate). Develop with butanol : acetic acid : water (7 : 1 : 2) to a height of 10–11 cm. Detect with Folin's reagent followed by a spray with 10% sodium carbonate. R_f ~0.47.
Source: Synthesized by catalytic hydrogenation of α-benzylamino-4-hydroxy-3-methoxyacetophenone,[4] and other methods.[4,5]
Storage: Protect from light and air.

References

1. T. Kappe and M. D. Armstrong, *J. Med. Chem.*, **8**, 368 (1965).
2. Z. Kahane and P. Vestergaard, *J. Lab. Clin. Med.*, **70**, 333 (1967).
3. *The U.S. Pharmacopeia*, XIX Revision (1975), p. 616.
4. G. Fodor, O. Kovacs, and T. Mecher, *Acta Chim. Acad. Sci. Hung.*, **1**, 395 (1951).
5. R. A. Heacock and O. Hutzinger, *Chem. Ind. (London)*, 595 (1961).
6. J. Axelrod, S. Senoh, and B. Witkop, *J. Biol. Chem.*, **233**, 697 (1958).

Additional Reference

D. J. Edwards and K. Blau, *Anal. Biochem.*, **45**, 387 (1972); gas-liquid chromatography of derivative.

BA-21
DL-Octopamine hydrochloride
IUPAC: (±)-α-(Aminoethyl)-p-hydroxybenzyl alcohol hydrochloride

C.A.: (±)-α-(Aminoethyl)-4-hydroxybenzenemethanol hydrochloride

Other: (norsynephrine hydrochloride, norsympatol hydrochloride)

Formula: $C_8H_{11}ClNO_2$
Formula Wt.: 188.6

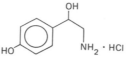

Description:

Appearance: Colorless crystalline powder.

Melting Point: ~168° (dec.).

Solubility: Soluble in water.

Stability: Aqueous solutions exposed to light slowly darken.

Absorption Spectrum: In 0.1 M HCl, λ_{max} 223 nm, ϵ ~8,550; λ_{max} 273 nm, ϵ ~1,280; λ_{sh} 279 nm.[1]

Fluorescence Spectrum: In water, λ_{ex} 279 nm, λ_{em} 323 nm. In 0.05 M sodium acetate buffer, pH 4, λ_{ex} 279 nm, λ_{em} 327 nm.

Reaction Product Fluorescence:[2] Reaction with 1-nitroso-2-napthol, λ_{ex} 465 nm, λ_{em} 553 nm.

Anion Test: Positive for chloride.[3]

Homogeneity: Thin-layer Chromatography

System 1. Spot 4 μl of a 0.5% solution in water on silica gel F254 Merck (precoated plate). Develop with butanol : acetic acid : water (7:1:2) to a height of 10–11 cm. Detect with Folin's reagent followed by a spray with 10% sodium carbonate. R_f~0.50.

System 2. Spot 4 μl of a 0.5% solution in water on cellulose F Merck (pre-coated plate). Develop with butanol : acetic acid : water (7:1:2) to a height of 10–11 cm. Detect with Folin's reagent followed by a spray with 10% sodium carbonate. R_f~0.50.

Source: Synthesized by reduction of 4-hydroxymandelonitrile with lithium aluminum hydride, or catalytic hydrogenation of an acetophenone precursor.[4] Resolution to D-(−)-octopamine and L-(+)-octopamine is accomplished with (+)-camphor-10-sulfonic acid.[5]

Storage: Protect from light and air.

References

1. T. Kappe and M. D. Armstrong, *J. Med. Chem.*, **8**, 368 (1965); report shoulder absorption.
2. S. Spector, K. Melmon, W. Lovenberg, and A. Sjoerdsma, *J. Pharmacol. Exp. Ther.*, **140**, 229 (1963); reaction product fluorescence method used.
3. *The U.S. Pharmacopeia*, XIX Revision (1975), p. 616.
4. M. Asscher, U.S. 2,585,988 (1952); *Chem. Abstr.*, **46**, 8150 (1952).
5. T. Kappe and M. D. Armstrong, *J. Med. Chem.*, **7**, 569 (1964).

Additional Reference

T. L. Perry and W. A. Schroeder, *J. Chromatogr.*, **12**, 358 (1963); ion exchange and paper chromatography.

BA-22
Phenethylamine hydrochloride
IUPAC: Phenethylamine hydrochloride

C.A.: Benzeneethanamine hydrochloride

Other: (β-phenethylamine hydrochloride,[1] 1-amino-2-phenylethane hydrochloride

Formula: $C_8H_{12}ClN$
Formula Wt.: 157.6

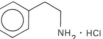

Description:

Appearance: Colorless platelets.

Melting Point: ~219–223°.

Solubility: Freely soluble in water. Recrystallized from ethanol.

Stability: The free amine absorbs carbon dioxide from the air.

Absorption Spectrum: In 0.1 M HCl, λ_{max} 206 nm, ϵ ~8,990; λ_{max} 257 nm, ϵ ~180.

Fluorescence Spectrum: In water, λ_{ex} 267 nm, λ_{em} 323 nm. In 0.05 M sodium acetate buffer, λ_{ex} 287 nm, ϵ ~373 nm.

Anion Test: Positive test for chloride.[2]

Homogeneity: Thin-layer Chromatography

System 1. Spot 4 μl of a 0.5% solution in water on silica gel F254 Merck (precoated plate). Develop with butanol : acetic acid : water (7:1:2) to a height of 10–11 cm. Detect with ninhydrin reagent followed by gentle heating. R_f ~0.56.

System 2. Spot 4 μl of a 0.5% solution in water on cellulose F Merck (precoated plate). Develop with butanol : acetic acid : water (7:1:2) to a height of 10–11 cm. Detect with ninhydrin reagent or with iodine vapor. R_f ~0.73.

Source: Catalytic hydrogenation of benzyl cyanide.[1]

Likely Impurities: Di-(β-phenylethyl)amine.

Storage: Protect from air.

References

1. *Org. Syn. Col. Vol.* III, 720 (1955).
2. *The U.S. Pharmacopeia*, XIX Revision (1975), p. 616.

Additional References

P. Cancalon and J. P. Klingman, *J. Chromatogr. Sci.*, **10**, 253 (1972); derivatives for gas–liquid chromatography.

D. J. Edwards and K. Blau, *Anal. Biochem.*, **45**, 387 (1972); gas–liquid chromatography.

H. Spatz and N. Spatz, *Biochem. Med.*, **6**, 1 (1972); reaction product fluorescence with p-dimethylaminocinnamaldehyde.

T. Nakajima, Y. Kakimoto, and I Sano, *J. Pharmacol. Exp. Ther.*, **143**, 319 (1964); paper chromatography, and ultraviolet and fluorescence spectra.

BA-23
Serotonin creatinine sulfate monohydrate
IUPAC: 3-(2-Aminoethyl)-1*H*-indol-5-ol creatinine
sulfate monohydrate
C.A.: 3-(2-Aminoethyl)-1*H*-indol-5-ol, compound with
2-amino-1,5-dihydro-1-methyl-4*H*-imidazole-4-
one (1:1) sulfate (1:1) (salt) monohydrate
Other: (5-hydroxytryptamine creatinine sulfate)

Formula: $C_{14}H_{23}N_5O_7S$
Formula Wt.: 405.4

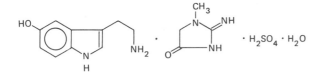

Water Content: 4.45%
Description:

Appearance: Colorless microcrystals; acceptable preparations
may be buff colored.

Melting Point: ~215° (dec.).

Solubility: In water, 20 mg/ml at 27°; 100 mg/ml at 50°. Insoluble
in most organic solvents.

Stability: Relatively stable in aqueous solutions under
anaerobic conditions. Unstable as the free base.[1]

Absorption Spectrum: In 0.1 M HCl, λ_{max} 219 nm, ϵ ~27,300;
λ_{max} 275 nm, ϵ ~5,800; λ_{sh} 293 nm.[2]

Fluorescence Spectrum: In water, λ_{ex} 295 nm, λ_{em} 352 nm. In
0.05 M sodium acetate buffer, λ_{ex} 295 nm, λ_{em} 352 nm.

Reaction Product Fluorescence: Reaction with *o*-phthal-
aldehyde in acid, λ_{ex} 340 nm, λ_{em} 463 nm.

Anion Test: Positive test for sulfate.[3]

Homogeneity: Thin-layer Chromatography

System 1. Spot 4 μl of a 0.5% solution on silica gel F254 Merck
(precoated plate). Develop with butanol : acetic acid : water
(7:1:2) to a height of 10–11 cm. Detect with Folin's reagent
followed by a spray of 10% sodium carbonate.

System 2. Spot 4 μl of a 0.5% solution on cellulose F Merck
(precoated plate). Develop with butanol : acetic acid : water
(7:1:2) to a height of 10–11 cm. Detect with a ninhydrin spray
followed by gentle heating.

	R_f	
	System 1.	*System 2.*
Serotonin	~0.53	~0.45
Creatinine	~0.21	~0.35

Source: Catalytic hydrogenation of 5-benzyloxytryptamine.[4–6]
Storage: Protect from light and air.

References

1. L. M. Morozovskaya, G. C. Tsvetkova, N. N. Suvorov, I. S. Tubina, E. I.
Mikhailovskaya, and A. A. Chemerisskaya, *Zh. Vses, Khim. Ova.*, **11**, 477 (1966); the
free base, m.p. 150–150.5°, is soluble in ethanol, but almost insoluble in water.
2. M. B. Zucker, B. K. Friedman, and M. M. Rapport, *Proc. Soc. Exptl. Biol. Med.*, **85**,
282 (1954); and M. M. Rapport, *J. Biol. Chem.*, **180**, 961 (1949); report shoulder
absorption in acid at 293 nm.
3. *The U.S. Pharmacopeia*, XIX Revision (1975), p. 617.
4. M. E. Speeter, R. V. Heinzelman, and D. I. Weisblat, *J. Am. Chem. Soc.*, **73**, 5514
(1951).
5. K. E. Hamlin and F. E. Fischer, *J. Am. Chem. Soc.*, **73**, 5007 (1951); report the
hydrochloride salt of 5-hydroxytryptamine to be hygroscopic and light sensitive.
6. L. Bretherick, K. Gaimster, and W. R. Wragg, *J. Chem. Soc.*, 2919 (1961).

Additional References

D. Keglevic and L. Stancic, *J. Labeled Compd.*, **3**, 144 (1967); and D. Keglevic-Brovet, S.
Kveder, and S. Iskric, *Croat. Chem. Acta*, **29**, 351 (1957); for the synthesis of carbon-14
side chain and ring labeled 5-hydroxytryptamine.
F. Cattabeni, S. H. Koslow, and E. Costa, *Science*, **178**, 166 (1972); derivatization for
gas–chromatographic mass spectrometric assay.

BA-24
Serotonin hydrogen oxalate
IUPAC: 3-(2-Aminoethyl)-1*H*-indol-5-ol oxalate
C.A.: 3-(2-Aminoethyl)-1*H*-indol-5-ol ethanedioate
(1:1) (salt)
Other: (5-hydroxytryptamine oxalate)

Formula: $C_{12}H_{14}N_2O_5$
Formula Wt.: 266.3

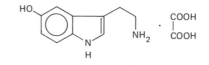

Description:

Appearance: Colorless needles.

Melting Point: ~198–199° (dec.).

Solubility: Soluble in water, insoluble in most organic solvents.

Stability: Relatively stable in aqueous solutions under
anaerobic conditions.

Absorption Spectrum: In 0.1 M HCl, λ_{max} 220 nm, ϵ ~19,100;
λ_{max} 276 nm, ϵ ~5,020; λ_{sh} 296 nm.[1]

Fluorescence Spectrum: In water, λ_{ex} 295 nm, λ_{em} 352 nm. In
0.05 M sodium acetate buffer, λ_{em} 295 nm, λ_{em} 352 nm.

Reaction Product Fluorescence: Reaction with o-phthalaldehyde in acid, λ_{ex} 340 nm, λ_{em} 463 nm.

Anion Test: Positive for oxalic acid.[2]

Homogeneity: Thin-layer Chromatography

System 1. Spot 4 μl of a 0.5% solution in water on silica gel
F254 Merck (precoated plate). Develop with butanol : acetic
acid : water (7:1:2) to a height of 10–11 cm. Detect with Folin's
reagent followed by a spray of 10% sodium carbonate. R_f ~0.53.

System 2. Spot 4 μl of a 0.5% solution in water on cellulose F
Merck (precoated plate). Develop with butanol : acetic acid : water (7:1:2) to a height of 10–11 cm. Detect with Folin's reagent
followed by a spray of 10% sodium carbonate. R_f ~0.43.

Source: Catalytic hydrogenation of 5-benzyloxytryptamine.[3,4]
Likely Impurities: Possible alkylamine salts, ammonium salt, or
bis-serotonin oxalate.
Storage: Protect from light, air, and moisture.

References

1. M. Bakri and J. R. Carlson, *Anal. Biochem.*, **34**, 46 (1970); report shoulder absorption
at 296 nm.
2. *The U.S. Pharmacopeia*, XIX Revision (1975), p. 616.
3. J. Harley-Mason and A. H. Jackson, *J. Chem. Soc.*, 1165 (1954).
4. C. D. Nenitzescu and D. Raileanu, *Chem. Ber*, **91**, 1141 (1958).

Additional References

R. P. Maickel and F. P. Miller, *Anal. Chem.*, **38**, 1937 (1966); reaction product fluores-
cence with o-phthalaldehyde.
H. Weissbach, *Std. Methods Clin. Chem.*, **4**, 197 (1963), D. Seligson (ed.), Academic
Press, NY; fluorescence assay for serotonin.

J. T. Vanable, *Anal. Biochem.*, **6**, 393 (1963); reaction product fluorescence with ninhydrin.

S. Udenfriend and H. Weissbach, *Methods Enzym.*, **6**, 598 (1963); paper chromatography and fluorometric assay.

T. Nogradi, P. D. Hrdina, and G. M. Ling, *Mol. Pharmacol.*, **8**, 565 (1974); nuclear magnetic resonance spectrum.

BA-25
DL-Synephrine hydrochloride
IUPAC: (±)-*p*-Hydroxy-α-[(methylamino)methyl] benzyl alcohol hydrochloride

C.A.: (±)-4-Hydroxy-α-[(methylamino)methyl] benzenemethanol hydrochloride

Formula: $C_9H_{14}ClNO_2$
Formula Wt.: 203.7

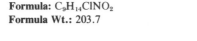

Description:
Appearance: Colorless crystals.
Melting Point: ~150–151°.
Solubility: Soluble in water.
Stability: Crystalline product is stable in air and light.
Absorption Spectrum: In 0.1 M HCl, λ_{max} 223 nm, ϵ ~7,360; λ_{max} 273 nm, ϵ ~1,080; λ_{sh} 279 nm.[1]
Fluorescence Spectrum: In water, λ_{ex} 267 nm, λ_{em} 332 nm. In 0.05 M sodium acetate buffer, pH 4.0, λ_{ex} 261 nm, λ_{em} 406 nm.
Reaction Product Fluorescence:[2] Reaction with 1-nitroso-2-napthol, λ_{ex} 465 nm, λ_{em} 553 nm.
Anion Test: Positive test for chloride.[3]
Homogeneity: Thin-layer Chromatography
System 1. Spot 4 μl of a 0.5% solution in water on silica gel F254 Merck (precoated plate). Develop with butanol : acetic acid : water (7:1:2) to a height of 10–11 cm. Detect with Folin's reagent followed by a spray with 10% sodium carbonate. R_f ~0.45.
System 2. Spot 4 μl of a 0.5% solution in water on cellulose F Merck (precoated plate). Develop with butanol : acetic acid : water (7:1:2) to a height of 10–11 cm. Detect with Folin's reagent followed by a spray with 10% sodium carbonate. R_f ~0.56.
Source: Catalytic hydrogenation of ω-methylamino-4-hydroxyacetophenone.[4,5] Resolution with bromocamphor-sulfonic acid.[6]
Storage: Protect from light and air.

References

1. T. Kappe and M. D. Armstrong, *J. Med. Chem.*, **8**, 368 (1965); report shoulder absorption.
2. S. Spector, K. Melmon, W. Lovenberg, and A. Sjoerdsma, *J. Pharmacol. Exp. Ther.*, **140**, 229 (1963); reaction product fluorescence method used.
3. *The U.S. Pharmacopeia*, XIX Revision (1975), p. 616.
4. M. Asscher, U.S. 2,585,988 (1952); *Chem. Abstr.*, **46**, 8150 (1952).
5. Ger. pat. 566,578 (1931); *Chem. Abstr.*, **27**, 2455 (1933).
6. H. Legerlotz, Ger. pat. 543,529 (1929); *Chem. Abstr.*, **26**, 3266 (1932).

BA-26
DL-Synephrine (+)-tartrate
IUPAC: (±)-*p*-Hydroxy-α-[(methylamino)methyl] benzyl alcohol (+)-tartrate (2:1) (salt)

C.A.: (±)-4-Hydroxy-α-[(methylamino)methyl]benzenemethanol[*R*-(*R**,*R**)]-2,3-dihydroxybutanedioate (2:1) (salt)

Other: (Oxedrine tartrate, Sympatol, Synthenate)

Formula: $(C_9H_{13}NO_2)_2 \cdot C_4H_6O_6$
Formula Wt.: 484.5

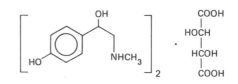

Description:
Appearance: Colorless crystals.
Melting Point: ~188–189° (dec.).
Solubility: Soluble in water.
Stability: Crystalline product is fairly stable in air and light.
Absorption Spectrum: In 0.1 M HCl, λ_{max} 223 nm, ϵ ~18,000; λ_{max} 273 nm, ϵ ~2,760.
Fluorescence Spectrum: In water, λ_{ex} 267 nm and 291 nm, λ_{em} 337 nm. In 0.05 M sodium acetate buffer, pH 4.0, λ_{ex} 280 nm, λ_{em} 331 nm.
Anion Test: Positive test for tartrate.[1]
Homogeneity: Thin-layer Chromatography
System 1. Spot 4 μl of a 0.5% solution in water on silica gel F254 Merck (precoated plate). Develop with butanol : acetic acid : water (7:1:2) to a height of 10–11 cm. Detect with Folin's reagent followed by a spray with 10% sodium carbonate. R_f ~0.44.
System 2. Spot 4 μl of a 0.5% solution in water on cellulose F Merck (precoated plate). Develop with butanol : acetic acid : water (7:1:2) to a height of 10–11 cm. Detect with Folin's reagent followed by a spray with 10% sodium carbonate. R_f ~0.57.
Source: Synthesized.[2,3]
Storage: Protect from light and air.

References

1. *The U.S. Pharmacopeia*, XIX Revision (1975), p. 617.
2. M. Asscher, U.S. 2,585,988 (1952); *Chem. Abstr.*, **46**, 8150 (1952).
3. Ger. pat. 566,578 (1931), to C. H. Boehringer Sohn A.-G.; *Chem. Abstr.*, **27**, 2455 (1933).

Additional References

S. Udenfriend, D. E. Dugan, B. M. Vasta, and B. B. Brodie, *J. Pharmacol. Exp. Ther.*, **120**, 26 (1957); fluorescence spectrum at pH 1.

T. L. Perry and W. A. Schroeder, *J. Chromatogr.*, **12**, 358 (1963); ion exchange and paper chromatography.

BA-27
Tryptamine
IUPAC: 3-(2-Aminoethyl)indole
C.A.: 1*H*-Indole-3-ethanamine

Formula: $C_{10}H_{12}N_2$
Formula Wt.: 160.2

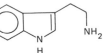

Description:

Appearance: Colorless needles or prisms; acceptable preparations may be buff colored.

Melting Point: ~116–118°.

Solubility: Soluble in ethanol, relatively insoluble in water and most of the other organic solvents. Recrystallized from ether or petroleum ether.

Absorption Spectrum: In 0.01 *M* HCl, λ_{max} 219 nm, ϵ ~33,900; λ_{max} 279 nm, ϵ ~5,580.

Fluorescence Spectrum: In 0.05 *M* sodium acetate buffer, λ_{ex} 280 nm, λ_{em} 373 nm.

Reaction Product Fluorescence: Reaction with *o*-phthalaldehyde, λ_{ex} 339 nm, λ_{em} 463 nm.

Homogeneity: Thin-layer Chromatography

System 1. Spot 4 μl of a 0.5% solution in water–methanol on silica gel F254 Merck (precoated plate). Develop with butanol : acetic acid : water (7 : 1 : 2) to a height of 10–11 cm. Detect with ninhydrin reagent followed by gentle heating or with Ehrlich's reagent. R_f ~0.55.

System 2. Spot 4 μl of a 0.05% solution in water–methanol on cellulose F Merck (precoated plate). Develop with butanol : acetic acid : water (7 : 1 : 2) to a height of 10–11 cm. Detect with ninhydrin reagent or with iodine in ethanol. R_f ~0.70.

Source: Reduction of 3-(2-nitrovinyl)indole, 3-indolylglyoxamide, or 3-indoleacetonitrile.[2,3]

Likely Impurities: Impurities may arise from oxidation by air at the 2-indolyl position and at the amine group.

Storage: Protect from light and air.

References

1. A. H. Jackson and A. E. Smith, *J. Chem. Soc.*, 3498 (1965).
2. R. J. Sundberg, *The Chemistry of Indoles*, 489 pp., Academic Press, NY (1970).
3. J. E. Saxton, in *The Indole Alkaloids*, R. H. F. Manske and H. L. Holmes (eds.), Vol. 8, pp. 8–10, Academic Press, NY (1965).

Additional References

A. H. Jackson and A. E. Smith, *J. Chem. Soc.*, 5510 (1964); protonation in acid media, ultraviolet and nuclear magnetic resonance spectra.

M. Bakri and J. R. Carlson, *Anal. Biochem.* **34**, 46 (1970); ultraviolet spectra of tryptamine and tryptophane metabolites.

R. P. Maickel and F. P. Miller, *Anal. Chem.*, **38**, 1937 (1966); reaction product fluorescence.

L. A. Cohen, J. W. Daly, H. Kny, and B. Witkop, *J. Am. Chem. Soc.*, **82**, 2184 (1960); NMR spectrum.

Y. Kanaoka, Y. Ban, T. Oishi, O. Yonemitsu, M. Terashima, T. Kimura, and M. Nakagawa, *Chem. Pharm. Bull.*, **8**, 294 (1960); infrared spectrum.

BA-28
Tryptamine hydrochloride
IUPAC: 3-(2-Aminoethyl)indole hydrochloride
C.A.: 1*H*-Indole-3-ethanamine hydrochloride

Formula: $C_{10}H_{13}ClN_2$
Formula Wt.: 196.7

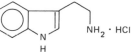

Description:

Appearance: Colorless needles.

Melting Point: ~253–256°.

Solubility: Moderately soluble in water, insoluble in most of the other organic solvents. Recrystallized from methanol.

Stability: Relatively stable in water when protected from oxidants. Sensitive to acids.

Absorption Spectrum: In 0.1 *M* HCl, λ_{max} 218 nm, ϵ ~33,600; λ_{max} 279 nm, ϵ ~5,520.

Fluorescence Spectrum: In water, λ_{ex} 283 nm, λ_{em} 376 nm. In 0.05 *M* sodium acetate buffer, λ_{ex} 283 nm, λ_{em} 366 nm.

Reaction Product Fluorescence: Condensation with formaldehyde, followed by oxidation with hydrogen peroxide to form Norharman;[1] λ_{ex} 365 nm, λ_{em} 440 nm.

Anion Test: Positive test for chloride ion.[2]

Homogeneity: Thin-layer Chromatography

System 1. Spot 4 μl of a 0.5% solution in water on silica gel F254 Merck (precoated plate). Develop with butanol : acetic acid : water (7 : 1 : 2) to a height of 10–11 cm. Detect with ninhydrin reagent followed by gentle heating, or with ultraviolet light. R_f ~0.56.

System 2. Spot 4 μl of a 0.5% solution in water on cellulose F Merck (precoated plate). Develop with butanol : acetic acid : water (7 : 1 : 2) to a height of 10–11 cm. Detect with ninhydrin reagent, or with iodine vapor. R_f ~0.70.

Source: Chemical reduction of 3-(2-nitrovinyl)indole[3] or 3-indolylglyoxamide[4,5] with lithium aluminum hydride, or by low-pressure catalytic hydrogenation of 3-indoleacetonitrile.[6]

Storage: Protect from light and air.

References

1. Data reported by a) S. M. Hess and S. Udenfriend, *J. Pharmacol. Exp. Ther.*, **127**, 175 (1960); b) S. M. Hess, *Methods Med. Res.*, **9**, 175 (1961).
2. *The U.S. Pharmacopeia*, XIX Revision (1975), p. 616.
3. E. H. P. Young, *J. Chem. Soc.*, 3493 (1958).
4. M. E. Speeter and W. C. Anthony, *J. Am. Chem. Soc.*, **76**, 6208 (1954).
5. F. V. Brutcher and W. D. Vanderwerff, *J. Org. Chem.*, **23**, 146 (1958).
6. M. Freifelder, *J. Am. Chem. Soc.*, **82**, 2386 (1960).

Additional References

W. E. Noland and P. J. Hartman, *J. Am. Chem. Soc.*, **76**, 3227 (1954); spectral constants.

J. A. Oates, *Methods Med. Res.*, **9**, 169 (1961); assay by fluorescence spectrum.

BA-29
Tyramine
IUPAC: *p*-(2-Aminoethyl)phenol
C.A.: 4-(2-Aminoethyl)phenol
Other: (*p*-hydroxyphenethylamine, tyrosamine)

Formula: $C_8H_{11}NO$
Formula Wt.: 137.2

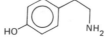

Description:

Appearance: Colorless crystals.
Melting Point: ~159–162°.
Solubility: One gram dissolves in 95 ml of water at 15°, or 10 ml of ethanol at the boiling point. Sparingly soluble in benzene and other organic solvents.
Absorption Spectrum: In 0.1 *M* HCl, λ_{max} 221 nm, ϵ ~7,750; λ_{max} 275 nm, ϵ ~1,730; λ_{sh} 280 nm.
Fluorescence Spectrum: In 0.05 *M* sodium acetate buffer at pH 4.0, λ_{ex} 281 nm, λ_{em} 323 nm.

Homogeneity: Thin-layer Chromatography

System 1. Spot 4 μl of a 0.5% solution in water–methanol on silica gel F254 Merck (precoated plate). Develop with butanol : acetic acid : water (7:1:2) to a height of 10–11 cm. Detect with ninhydrin reagent followed by gentle heating. R_f ~0.50.

System 2. Spot 4 μl of a 0.5% solution in water–methanol on cellulose F Merck (precoated plate). Develop with butanol : acetic acid : water (7:1:2) to a height of 10–11 cm. Detect with ninhydrin reagent. R_f ~0.61.

Source: Synthetic.[2-4]
Storage: Protect from air.

References

1. T. Kappe and M. D. Armstrong, *J. Med. Chem.*, **8**, 368 (1965), report shoulder absorption.
2. J. J. Buck, *J. Am. Chem. Soc.*, **55**, 2593, 3388 (1933); catalytic reduction of 4-hydroxymandelonitrile.
3. E. Waser, *Helv. Chim. Acta*, **8**, 758 (1925); from tyrosine.
4. K. Dose, *Chem. Ber.*, **90**, 1251 (1958); decarbonylation of tyrosine.

Additional References

E. G. McGeer and W. H. Clark, *J. Chromatogr.*, **14**, 107 (1964); paper chromatography.
T. L. Perry and W. A. Schroeder, *J. Chromatogr.*, **12**, 358 (1963); ion exchange and paper chromatography.
A. Vahida and D. V. Sankar, *J. Chromatogr.*, **43**, 135 (1969); paper chromatography and partition thin-layer chromatography.
J. Lundstrom and S. Agurell, *J. Chromatogr.*, **30**, 271 (1967); thin-layer chromatography.

BA-30
Tyramine hydrochloride
IUPAC: *p*-(2-Aminoethyl)phenol hydrochloride
C.A.: 4-(2-Aminoethyl)phenol hydrochloride
Other: (*p*-hydroxyphenethylamine hydrochloride)

Formula: $C_8H_{12}ClNO$
Formula Wt.: 173.6

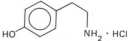

Description:

Appearance: Colorless crystals.
Melting Point: ~272–275°.
Solubility: Soluble in water and alcohol.
Absorption Spectrum: In 0.1 *M* HCl, λ_{max} 222 nm, ϵ ~7,570; λ_{max} 275 nm, ϵ ~1,560.
Fluorescence Spectrum: In water, λ_{ex} 279 nm, λ_{em} 332 nm. In 0.05 *M* sodium acetate buffer at pH 4.0, λ_{ex} 279 nm, λ_{em} 323 nm.
Reaction Product Fluorescence: Reaction with 1-nitroso-2-napthol,[1] λ_{ex} 465 nm, λ_{em} 553 nm.
Anion Test: Positive test for chloride.[2]

Homogeneity: Thin-layer Chromatography

System 1. Spot 4 μl of a 0.5% solution in water on silica gel F254 Merck (precoated plate). Develop with butanol : acetic acid : water (7:1:2) to a height of 10–11 cm. Detect with spray of ninhydrin reagent followed by gentle heating. R_f ~0.51.

System 2. Spot 4 μl of a 0.5% solution in water in cellulose F Merck (precoated plate). Develop with butanol : acetic acid : water (7:1:2) to a height of 10–11 cm. Detect with ninhydrin reagent. R_f ~0.57.

Source: Synthetic (see Tyramine).
Storage: Protect from light and air.

References

1. S. Spector, K. Melmon, W. Lovenborg, and A. Sjoerdsma, *J. Pharmacol. Exp. Ther.*, **140**, 229 (1963); reaction product fluorescence method.
2. *The U.S. Pharmacopeia*, XIX Revision (1975), p. 616.

Additional References

N. Seiler and M. Wieckmann, *Hoppe Seyler's Z. Physiol. Chem.*, **337**, 229 (1964); reaction product fluorescence with formaldehyde and ammonia.
R. A. Alcaron, *Arch. Biochem. Biophys.*, **113**, 281 (1960); reaction product fluorescence with acrolein and resorcinol.
P. Cancalon and J. D. Klingman, *J. Chromatogr. Sci.*, **10**, 253 (1972); derivatization for gas–liquid chromatography.
J. E. Sinsheimer and E. Smith, *J. Pharm. Sci.*, **52**, 1080 (1963); infrared and ultraviolet spectra of tetraphenylborate derivative.

ISBN 0-309-02601-6